Dahlem Workshop Reports
Life Sciences Research Report 20
Neuronal-glial Cell Interrelationships

The goal of this Dahlem Workshop is:
to examine neuronal-glial cell interrelationships
and their modification in disease
with particular reference to multiple sclerosis

Life Sciences Research Reports
Editor: Silke Bernhard

Held and published on behalf of the
Stifterverband für die Deutsche Wissenschaft

Sponsored by:
Hermann and Lilly Schilling-Stiftung
für medizinische Forschung
im Stifterverband für die Deutsche Wissenschaft
Senat der Stadt Berlin

Neuronal-glial
Cell Interrelationships

T. A. Sears, Editor

Report of the Dahlem Workshop on
Neuronal-glial Cell Interrelationships:
Ontogeny, Maintenance, Injury, Repair
Berlin 1980, November 30 – December 5

Q H
359
N 494
1980

Rapporteurs:
R. L. Barchi · G. R. Strichartz · P. A. Walicke · H. L. Weiner

Program Advisory Committee:
T. A. Sears, Chairman · A. J. Aguayo · B. G. W. Arnason
H. J. Bauer · B. N. Fields · W. I. McDonald · J. G. Nicholls

Springer-Verlag Berlin Heidelberg New York 1982

Copy Editor: M. Cervantes-Waldmann
Photographs: E. P. Thonke

With 5 photographs, 13 figures, and 8 tables

ISBN 3-540-11329-0 Springer-Verlag Berlin Heidelberg New York
ISBN 0-387-11329-0 Springer-Verlag New York Heidelberg Berlin

CIP-Kurztitelaufnahme der Deutschen Bibliothek:
Neuronal-glial cell interrelationships: report of the Dahlem Workshop on
Neuronal-glial Cell Interrelationships: Ontogeny, Maintenance, Injury, Repair,
Berlin 1980, November 30 – December 5 / T. A. Sears, ed. Rapporteurs: R. L. Barchi . . .
[Dahlem Konferenzen. Held and publ. on behalf of the Stifterverb. für d. Dt. Wiss.
Sponsored by: Hermann-and-Lilly-Schilling-Stiftung für Med. Forschung im
Stifterverb. für d. Dt. Wiss.; Senat d. Stadt Berlin]. – Berlin; Heidelberg; New York:
Springer, 1982.
    (Life sciences research report; 20)
    (Dahlem workshop reports)
NE: Sears, Thomas A. [Hrsg.]; Barchi, Robert L. [Mitverf.]; Workshop on
Neuronal-glial Cell Interrelationships: Ontogeny, Maintenance, Injury, Repair
<1980, Berlin, West>; Dahlem Konferenzen; GT

© Dr. S. Bernhard, Dahlem Konferenzen, Berlin 1982
Printed in Germany

Printing: Proff GmbH & Co. KG, D-5340 Bad Honnef
Bookbinding: Graphischer Betrieb Konrad Triltsch, D-8700 Würzburg
2131/3014 – 5 4 3 2 1 0

Table of Contents

The Dahlem Konferenzen

DIRECTOR:
Silke Bernhard, M.D.

FOUNDATION:
Dahlem Konferenzen was founded in 1974 and is supported by the Stifterverband für die Deutsche Wissenschaft, the Science Foundation of German Industry, in cooperation with the Deutsche Forschungsgemeinschaft, the German Organization for Promoting Fundamental Research, and the Senate of the City of Berlin.

OBJECTIVES:
The task of Dahlem Konferenzen is:
   to promote the interdisciplinary exchange of scientific information and ideas,

   to stimulate international cooperation in research, and

   to develop and test different models conducive to more effective scientific meetings.

AIM:
Each Dahlem Workshop is designed to provide a survey of the present state of the art of the topic at hand as seen by the various disciplines concerned, to review new concepts and techniques, and to recommend directions for future research.

PROCEDURE:
Dahlem Konferenzen approaches internationally recognized scientists to suggest topics fulfilling these criteria and to propose members for a Program Advisory Committee, which is responsible for the workshop's scientific program. Once a year, the topic suggestions are submitted to a scientific board for approval.

TOPICS:
The workshop topics should be:
   of contemporary international interest,

   timely,

   interdisciplinary in nature, and

   problem-oriented.

PARTICIPANTS:
The number of participants is limited to 48 for each workshop,
and they are selected exclusively by a Program Advisory Committee.
Selection is based on international scientific reputation alone
and is independent of national considerations, although a balance between Europeans and Americans is desirable. Exception is
made for younger German scientists for whom 10% of the places
are reserved.

THE DAHLEM WORKSHOP MODEL:
A special workshop model has been developed by Dahlem Konferenzen,
the *Dahlem Workshop Model*. The main work of the workshop is done
in four small, interdisciplinary discussion groups, each with
12 members. Lectures are not given.

Some participants are asked to write background papers providing
a review of the field rather than a report on individual work.
These are circulated to all participants 4 weeks before the meeting with the request that the papers be read and questions on
them formulated *before* the workshop, thus providing the basis
for discussions.

During the workshop, each group prepares a Report reflecting the
essential points of its discussions, including suggestions for
future research needs. These reports are distributed to all
participants at the end of the workshop and are discussed in
plenum.

PUBLICATION:
The Dahlem Workshop Reports contain:
  the Chairperson's introduction,

  the background papers, and

  the Group Reports.

Stephen W. Kuffler

Stephen W. Kuffler

(1913-1980)

Time after time, in a career that spanned 40 years, Stephen
Kuffler made experiments on fresh topics, hitherto confused
or ignored, in which he revealed fundamental mechanisms and
laid paths for future research to follow. In each instance,
the striking feature of his work is the way in which the right
problem was tackled at the right time, using the right preparation. One can, for example, think of the studies on denervation supersensitivity, stretch receptors and muscle spin-
dles, efferent control, pre- and postsynaptic inhibition, GABA
and peptides as transmitters, integration in the retina, and
the detailed analysis of synaptic transmission.

In the context of this Dahlem meeting on neuronal-glial interactions and multiple sclerosis, nowhere are these characteristics more clearly demonstrated than in the pioneering studies
he made on neuroglial cells with David Potter at Harvard Medical
School in the early 1960s. The experiments arose as a result
of teaching a course on the nervous system to medical students,
during which it became clear that there was no hard information
available concerning the physiology of the cells that make up
the larger part of the mammalian brain. In those experiments,
the key question was to determine the membrane properties of
glial cells in contrast to those of the neurons they surround.
What were the resting potentials and what was their ionic basis?
Were the membranes of glial cells permeable to K, to Na, to Cl,
or to all small molecules and ions? Did glial cells give impulses? Did substances like K or Na move through glial cells
or around them by way of extracellular spaces? What signals
passed between the two types of cells? These well-defined,
manageable questions were in marked contrast to a spate of ro-
mantic speculations that were in vogue at the time, proposing
roles for glial cells in memory, learning, and bulk transport
of materials through the brain. At the same time it was by no
means obvious how one should record with microelectrodes from

the tiny glial cells that are inextricably mixed up with neurons. In a systematic search for the most suitable preparation, Kuffler and Potter found the leech, the nervous system of which contains large glial cells that made it possible to make direct, unambiguous and reliable measurements of their membrane properties. Later with Orkand, frogs and mudpuppies were shown to have glial cells with similar properties. It was then a relatively short step for others to examine the mammalian brain because now electrical criteria were available for recognizing the glial cells, which turned out to be remarkably similar to their counterparts in lower animals. While many problems still remain concerning the possible roles of glial cells for spatial buffering of potassium and for signalling to neurons, the paper by Kuffler and Potter remains the cornerstone (as Orkand's paper, this volume, clearly shows).

What gave each new paper by Stephen Kuffler that special quality which made it such a pleasure to read? Partly it was the unremittingly high standards of evidence, partly the elegance of the approach and the beautiful figures, and partly the suppressed excitement of wondering - what would he tackle next? In addition, most of the experiments combined high technical virtuosity with directness of approach and a clarity of thought matched by the style of the writing. Moreover, one knew that, right up to the end, he himself had done every experiment he described.

A second feature of Stephen Kuffler's work that seems particularly relevant to the Dahlem conferences organized by Silke Bernhard is the multidisciplinary approach. To this end, he, more than anyone else, gave meaning to the idea of Neurobiology - the discipline in which the nervous system is studied in terms of cell biology, using biochemical, physiological, immunological, and anatomical approaches. At Harvard he created, for the first time, a department in which he brought together people from widely different disciplines who collaborated and talked together and thereby allowed new ways of thinking to evolve. He

helped also to create interdisciplinary courses for young
scientists at Woods Hole. The list of his collaborators and
pupils is a "Who's Who" of leading scientists in their own
right who were nevertheless touched and inspired by their
contact with him.

The list of his personal attributes is too hard to describe
adequately. They were clearly in evidence at the Dahlem
Conference he attended in 1977. Those who knew him saw that
unique combination of tolerance and firmness, kindness without sentimentality, good sense with enduring humor, with jokes
and puns that often had an end but no beginning but still made
one laugh.

Long walks, long talks, relaxed meals, and quiet silences with
friends were among his pleasures and contributed to the lasting
indelible memories he gave his friends. Those memories are so
strong that the personal loss is perhaps mitigated by the priv-
ilege and gratitude for having had such a man in our time.

He was the John Franklin Enders University Professor at Harvard
and was closely associated with the Marine Biological Laboratory at Woods Hole. Among his many honors and distinctions,
he was a foreign member of the Royal Society.

J.G. Nicholls

Neuronal-glial Cell Interrelationships, ed. T.A. Sears, pp. 5-10.
Dahlem Konferenzen 1982. Berlin, Heidelberg, New York: Springer-Verlag.

Introduction

T. A. Sears
Sobell Dept. of Neurophysiology
Institute of Neurology
London WC1N 3BG, England

This volume is an account of a Workshop at which a small group
of basic and clinical scientists met to discuss the topic of
Neuronal-glial Cell Interrelationships and in particular, how
these relationships are affected by "Multiple Sclerosis."

Multiple Sclerosis (MS) is a neurological disease first recognized by Charcot in 1868. A summary of the salient clinical and
pathological features of the disease is given by W.I. McDonald
in the opening chapter but here, by way of introduction, I need
briefly to refer to them in order to reveal both the setting in
which the need for this Dahlem Workshop was conceived and the
factors which governed the shape of the meeting. MS is characterized by a variety of symptoms and signs, among which disturbances of sensation such as blurring of vision, numbness, and
tingling are most common, coupled with weakness, paralysis, disorders of gait, and incontinence, while spastic paresis may develop as a longer term complication. These pathophysiological
states occur in various combinations according to the idiosyn-
cratic dissemination of the underlying lesions throughout the
central nervous system. The lesions, which appear histologically
as "plaques," are seen where the sheaths of myelinated fibers
are lost with preservation of the axons and where a substantial
activation of astrocytes has led to local gliosis or scarring.

MS is thus a disease characterized by disordered neuronal-glial
cell interrelationships.

MS generally strikes first in early adult life but having
struck, often to be followed by a period of virtually complete
recovery, pursues a nearly unpredictable course of relapse and
remission that imposes considerable physiological and psychological burdens on its victims, while frustrating the physician's
wish to provide a hopeful, rather than condemning prognosis.
At the present time the cause of the disease is unknown, its
cure evades detection, and the treatment of its symptoms is em-
pirical and haphazardly effective, to say the least, notwithstanding the overall improvement which has occurred in the lot
of patients through better general management, physiotherapy,
and social rehabilitation, as emphasized by Bauer (1). No matter how depressing this state of affairs may seem to be, it must
not, however, be viewed too pessimistically. An enormous increase in MS orientated research by clinical and basic scientists
has occurred during the last decade with the result that new
knowledge and technologies have been brought to bear on the dual
goal first, of discovering the etiology of the disease, and second, of developing effective treatments for the disease itself,
as well as of its symptoms, based rationally on the underlying
anatomical, biochemical, and physiological pathology as studied
at the cellular level. Admittedly these goals have yet to be
realized, but sufferers of MS might gain some comfort from the
knowledge that the fight against this cruel disease is well and
truly joined.

Major difficulties face any research worker in this field, not
least due to the wide-ranging interdisciplinary nature of many
of the relevant problems and the corresponding diversity of the
methods needed for their solution. MS has no respect for the
boundaries within which, in the past, anatomists, physiologists,
pharmacologists, immunologists, and virologists, etc., have
worked, i.e., within the traditional, methodologically circumscribed disciplines which arose in relation to the structure of
medicine in the 19th century. Unfortunately, the ever-increasing

need for an interdisciplinary approach to research, although
scientifically desirable and laudable, is not easily met by the
individual investigator, a statement which I must now qualify
lest it be taken as a faint-hearted view of the problems which
confront us in this or any other field of disease-orientated research.

In recent years the growth and scope of MS research parallels,
in fact reflects, that which has occurred more generally concerning research at all levels of complexity into the nature and modes
of operation of the nervous systems of different animals. With
respect to these developments Cowan (2) has observed that "this
has led to the gradual emergence of a new, interdisciplinary approach to the study of the nervous system which has come to be
known as Neuroscience."

At the center of neuroscience stands man striving to comprehend
himself, not only in terms of the nuts and bolts of his own nervous system and that of lower animals, but perhaps preoccupied
most of all with the higher level nervous functions of perception,
volition, cognition, and mentation, which characterize his "self."
The investigation of these processes depends ultimately on research on man himself and the analysis of these processes in
depth often must wait on Nature's own experiments to provide,
through disease, the chance anatomical or biochemical lesions
which dissect human behavior and expose the residual functions
for scientific study. As William Harvey clearly stated in the
last of his preserved letters written in 1657, the year of his
death: "Nature is nowhere accustomed more openly to display her
secret mysteries than in cases where she shows traces of her
workings apart from the beaten path; nor is there any better way
to advance the proper practice of medicine than to give our minds
to the discovery of the usual law of nature, by careful investigation of cases of rarer forms of disease. For it has been found
in almost all things, that what they contain of useful or of applicable, is hardly perceived unless we are deprived of them, or
they become deranged in some way" (see (3)).

Harvey's prophetic remarks are nowhere more applicable than in
MS research. The body of evidence derived through the clinical
and laboratory study of this and other demyelinating diseases
has, and continues to play, a crucial role in the development of
concepts and the shaping of fundamental questions that have been
posed, for example, in relation to the effects of demyelination
on nervous transmission. More generally, the study of human
neurological disease, especially when lesions are focal in distribution or functionally circumscribed, is absolutely critical
to the testing of hypotheses which relate in particular to higher
level functions, so that clinical research in neurology and psy-
chology can contribute to a number of the disciplines embraced
by neuroscience, in addition to standing as disciplines within
their own right. From the above it will be recognized that the
spectrum of subjects encompassed by neuroscience is vast, ranging
from membrane biophysics to neuropsychology to instance but two
very different disciplines.

Notwithstanding the willingness and need for neuroscientists to
cross the boundaries of the traditional disciplines, the difficulties of so doing should not be underestimated. Each discipline
is highly specialized, it invariably requires the acquisition
of considerable manual and technical skills and, perhaps most
intimidating for those who seek detailed information about another discipline, is enshrouded in a dense web of complex terminology that easily tries the patience of all but the highly motivat-
ed. So the need to specialize in depth, yet at the same time to
maintain a broad perspective, imposes a considerable burden on
the individual, and consequently it is not surprising that an increasing number of scientists achieve their aims through collabo-
rative research. But this very growth of neuroscience, occasioned
by success, popular appeal, and mushrooming publication, poses
the further problem of effective communication between different
disciplines, especially between the basic and clinical scientists
working within them. Although in principle these two groups
should now interact more readily under the umbrella of neuroscience, I am not yet convinced that this is indeed the case,

even though our present Workshop should dispose of my doubts!
The sheer scale of the problem of achieving expertise within a
discipline is a difficult enough task, but the cultural and pro-
fessional factors which separate the career development and work
of clinicians and scientists limit the essential element of personal contact that provides the spur for successful collaboration
or the obtaining of expert advice. This makes it all the more
important for specialists within one field to continue to try to
understand the needs, problems, and terminology of related fields.
The ideal forum to promote such an exchange is the kind of Workshop organized by the Dahlem Konferenzen - but how can we best
structure one that could identify future areas of research which
would optimize the attack against MS?

Not surprisingly, most meetings on MS center around the clinical
problems and on the direct search in the tissues and fluids of
MS patients for clues that would identify immunological, virological, toxicological, or other factors as the etiologic agent(s).
This search would clearly be aided by an up-to-date knowledge of
the normal cellular mechanisms with which the disease process
might interfere. Furthermore, until the etiology of MS is understood, "activity" of the disease process itself rests virtually
entirely on the neurological assessment of disordered function,
although, as discussed by Arnason and Waksman (this volume), there
is accumulating evidence that the number of circulating suppressor
lymphocytes may provide reliable, independent evidence of disease
"activity." Be this as it may, clear understanding of the physiological abnormalities which occur in demyelination, and of the
kind of compensatory mechanisms which may be involved, is clearly
essential to the development and assessment of curative and
symptomatic therapies, whose efficacy can only be judged through
improved function.

With these and many other considerations in mind, an alternative
strategy evolved for this Workshop which was to base it on a discussion of normal neuronal-glial cell interrelationships, with MS
then considered as a process which interfers with these relationships. This approach, which places emphasis on cellular mechanisms,

allowed a small group of clinical scientists with expert knowledge
of different aspects of the disease to interact with a group of
basic scientists, some of whom were already directly concerned in
MS orientated research, while others, whose expertise was ur-
gently needed in this field of work, had yet to become involved.

MS expresses itself in young adults but evidence from epidemiological, genetic, and familial studies points to the possibility
of an infective agent operating in early childhood. It was appropriate, therefore, to separate our discussions of normal
neuronal-glial cell interrelationships into Ontogeny and Maintenance. The discussion of the third group occurred under the
heading Injury, in which consideration was specifically given to
the cellular mechanisms through which MS and known infectious and
toxic agents may operate and the functional consequences of injury. The fourth group dealt with the question of Repair, a particularly important problem bearing in mind the implications which
such discussion has for those already suffering from the disease.
The reader will discover some measure of overlap in the State of
the Art Reports, but this was important for it served to provide
a consensus opinion on important lines for future research, some
of which have already been implemented.

To conclude this introduction for the reader, all present at this
Workshop were deeply conscious of the then recent and untimely
death of Stephen Kuffler. His personal contribution to knowledge
of neuronal-glial interrelationships has no parallel, yet this was
but one of his major contributions to neuroscience, as can be understood from the tribute which John Nicholls has made in the dedica-
tion of this volume to a great scientist and truly lovable man.

REFERENCES

(1) Bauer, H.J. 1978. Problems of symptomatic therapy in Multi-
    ple Sclerosis. Neurology 28(2): 8-20.

(2) Cowan, H.M. 1978. Preface. Ann. Rev. Neurosci. 1: 5-7.

(3) D'Arcy Power 1898. William Harvey. In Masters of Medicine.
    London: T. Fisher Unwin.

Neuronal-glial Cell Interrelationships, ed. T.A. Sears, pp. 11-24.
Dahlem Konferenzen 1982. Berlin, Heidelberg, New York: Springer-Verlag.

Clinical Problems of Multiple Sclerosis

W. I. McDonald
University Dept. of Clinical Neurology
Institute of Neurology
London WC1N 3 BG, England

Abstract. Multiple sclerosis is a relatively common disorder of
the central nervous system characterized pathologically by focal
areas of myelin destruction with preservation of the axons and
clinically by a relapsing and remitting course with persistent
disability developing ultimately in a majority of patients. The
cause is unknown but both environmental and genetic factors appear to be important. The former is probably infective, possibly
viral in nature and the latter is probably related to the control
of immune regulation. There is evidence that the myelin breakdown is mediated immunologically, but the mode of interaction of
the environmental and genetic factors which leads to the development of the disease, the mechanism of precipitation of the relapses, and the mechanisms of remission are obscure. There is no
effective treatment.

INTRODUCTION

Multiple sclerosis (MS) is one of the more common primary diseases
of the central nervous system in developed countries. It is characterized by episodes of neurological disturbance usually lasting
for weeks or months which in the early stages of the illness tend
to remit. As the years go by the deficit left by each relapse
accumulates and about two-thirds of the patients ultimately enter
a phase of more or less steady progression. Approximately 50%
of the patients die within 20 years of onset. The cause is unknown and there is no effective treatment.

This broad outline conceals a number of important problems: the
peculiar geographical distribution and racial incidence, the

great variability in the course of the illness, the predilection of the lesions for certain sites in the nervous system,
the persistence of microscopic and physiological evidence of
lesions despite complete symptomatic recovery and the existence
of immunological abnormalities whose significance is unknown.
In this introductory paper, I shall outline the main characteristics of the disease and review briefly what is known of
its cause, emphasizing the fundamental questions which the illness raises.

PATHOLOGY (2,14)
There are two essential pathological characteristics of MS -
the lesions themselves (the plaques of demyelination which
ultimately become hard or sclerotic) and their multiplicity
and patchy dissemination throughout the central nervous system.
The two names commonly used for the disease (multiple and disseminated sclerosis) are derived from these characteristics.

The plaques are irregularly shaped and usually have a small
vein at the center although in very large lesions this primarily perivenous distribution may be obscured. Plaques vary
in length from a few hundred μm to several centimeters. In
advanced cases very little myelin may survive in large regions
of the central white matter. It is a curious and unexplained
characteristic of the disease that the plaques, while they may
occur anywhere in the central nervous system, have certain sites
of predilection - the optic nerves which are almost invariably
involved at some stage, the brain stem, the cervical spinal
cord, and the periventricular regions.

The time course of development of the lesion is not accurately
known, but the broad outline of events can be pieced together
from correlations between post-mortem changes and the clinical
history and from serial sectioning of individual lesions.

In the chronic plaque the dominant histological features are
demyelination - destruction of the myelin sheath with preservation of continuity of the majority of axons - and the fibrous

glial scar formed by the intense proliferation of the astrocytes.
The demyelination may extend for several millimeters but some
fibers show short (approximately 100 μm) lengths of full- or
partial-thickness myelin loss (18). Groups of demyelinated
axons may lie very close to each other, sometimes with their
axolemmas in contact but more often the axons are separated
from each other by fibrillary astrocytic processes (19). Although demyelination is the dominant feature of the established
MS lesion, there is always some Wallerian-type degeneration of
axons. Occasionally the loss is extensive. The reasons for
this variability are unknown. The oligodendrocytes (the myelin
forming cells) are absent from the lesion itself (14). Within
a minority of old plaques the blood vessels may be cuffed with
lymphocytes and phagocytic cells (2). Extension of such cuffs
beyond a chronic plaque is rare (2).

Active plaques are characterized not only by demyelination but
by hypercellularity at the edges, engorgement of veins, and
by evidence of lipid products resulting from the breakdown of
myelin (2). The nature of the cells at the edge of the plaque
is not settled but some at least are astrocytes. In some plaques
(usually those with intense mononuclear perivascular cuffing)
there is marked edema at the edges of the lesion, some at least
of which is contained within the astrocytes.

Perivascular cuffing is prominent in relation to the active
lesion. Predominantly lymphocytic cuffs are found at the mar-
gin of the lesion extending into apparently normal white matter,
but cuffs composed chiefly of plasma cells are usually found
within it. These observations have led to the suggestion that
the active plaque extends along the venules. Larger plaques
also appear to extend centrifugally and evidence of recent
myelin breakdown in the absence of much lymphocytic infiltration is sometimes found at the edges of old lesions (19).

The mechanism of demyelination is still not fully understood.
It appears to depend upon microglial cells coming into contact

with the myelin sheath (19). Phagocytic stripping of myelin
such as has been observed in a variety of experimental lesions
has not been reported.

Biochemical studies of active plaques have shown that lysozome
hydrolase activity is increased and that there is selective
loss of myelin basic protein with the appearance of lower molecular weight peptides. The cellular origin of the enzymes has
not been established but there is evidence that some at least
may come from macrophages (6).

Whatever the cellular mechanism of demyelination in the MS
plaque, there is good evidence that the process is mediated
immunologically. The presence of plasma cells in the active
lesion has been mentioned already. Extra-cellular IgG can be
demonstrated in both recent and old plaques and the immunoperoxidase method has shown that intracellular IgG is present
in the mononuclear perivascular cuffs in recent plaques (9).
An increase in oligoclonal IgG in the CSF is an almost constant
finding in MS (22). Exactly what happens in the immunological
reaction and what triggers it are uncertain. Antel et al. (3)
however, have recently provided good evidence for changes in
lymphocyte subpopulations during exacerbations of MS.

The question of whether remyelination occurs in MS has long been
debated, but the first good evidence that it does so is very recent (20). It is however limited, being confined to the edges
of the lesion and involving only a tiny minority of the fibers
in the plaque. The new myelin is central in type implying an
oligodendrocyte origin, but in MS, as in a variety of experimental demyelinating lesions, a few Schwann cells may invade
the CNS and from new myelin. There is little data available
concerning the dimensions of the new myelin sheaths, but what
there is suggests that (as in experimental lesions) they are
inappropriately thin and short for the diameter of the axons.
Whether normal dimensions can ever be restored is unknown, but
it is clear that in most MS lesions at postmortem the great
majority of axons remain demyelinated.

CLINICAL FEATURES (12,15,17)

The patient with symptoms of multiple sclerosis usually first
presents in his or her twenties or thirties. Cases presenting
before the age of 10 or after the age of 60 are extremely rare.
Females are rather more frequently affected than males (F:M,
1.5:1).

The symptoms are broadly of two kinds - positive (e.g., tingling)
and negative (e.g., weakness). About two-thirds of patients
present with sensory symptoms (approximately one-third having
visual loss and one-third tingling or numbness) and about 40%
with weakness in the limbs or disturbances of gait.

The usual course of events in the first few episodes is that
the symptoms develop over a matter of a few days or a week or
two, persist for several weeks and gradually subside over a
month or two. In early attacks, full symptomatic recovery is
usual and indeed physical examination may reveal no residual
deficit.

Further episodes of neurological disturbance occur after widely
varying intervals. At one extreme there is a small group of
patients, often in their early twenties, who have several relapses a year and in whom the illness is fatal within about
five years. At the other extreme there are patients who go for
20 or 30 years or even longer between the first and second episodes. There are a number of reports of patients with lesions
characteristic of multiple sclerosis being found at postmortem
in individuals who had no recorded neurological symptoms during
life. Many patients have a number of symptom-free years before
having a series of relapses. A permanent neurological deficit
then gradually accumulates. In a small proportion of cases
(10-20% of different series) the illness is steadily progressive
from the onset. This type of course tends to be more frequent
in patients with a later age of onset.

MS is rarely a direct cause of death. When it is a contributory
cause, it usually acts by imposing immobility and predisposing
to infection and impaired renal function.

It is obvious from this account that the course of the illness
varies greatly. The factors influencing it are poorly understood (15). There is good evidence that relapses may be precipitated occasionally by vaccination against smallpox and less
frequently after other types of inoculations. There is some
statistical evidence that relapses may be precipitated by surgi-
cal operations, injury, infections, and in the puerperium, but
the frequency with which relapses follow any of these events
is low and in most patients there is no obvious event with which
the appearance of new symptoms can be associated.

DIAGNOSIS
There is no specific test for MS and the diagnosis remains
clinical (16). A diagnosis of "clinically definite multiple
sclerosis" depends on the demonstration of unequivocal evidence
of abnormalities in at least two necessarily separate sites in
the central white matter in a patient with a history of at least
two episodes of neurological disturbance. Support for the diagnosis comes from finding an oligoclonal pattern in the gamma
globulins at CSF electrophoresis. This change is present in
90% of clinically definite cases but is not specific, being
found in other diseases in which there is evidence of an immunological reaction in relation to the nervous system, such as acute
idiopathic polyneuritis, chronic meningitis, sarcoidosis, and
neurosyphillis (22).

One of the difficulties in diagnosis - that of identifying a
second necessarily separate lesion - has been overcome recently
by the application of evoked potential recording techniques to
patients with neurological disease. In the visual, auditory,
and somatosensory systems, it is easy to identify abnormalities
in the evoked potentials even in the absence of corresponding
deficits on physical examination (16). The visual evoked potential is the most useful of these tests in the clinic being
abnormal in about 80% of patients with clinically definite MS,
including more than 50% of those with normal visual function.
Evoked potential abnormalities are not specific to MS but are

useful as a means of detecting the presence of a lesion, the
significance of which must be interpreted in the light of other
data.

ETIOLOGY
There is now good evidence that there are two major factors involved in the causation of multiple sclerosis - environmental
and genetic factor.

Environmental Factors
The evidence for the existence of an environmental factor is
epidemiological (1,15). Dean originally showed that the prevalence of multiple sclerosis (the number of patients with the
disease on a specified date) in the population of Northern
European origin born in South Africa and in adults who had
migrated there in childhood was low, whereas among those who had
migrated there in adult life, it was much higher and was similar to that found in their country of origin. The effects of
migration and of age of migration on prevalence have been confirmed in Israel (12). It is now clear that there is a general
north/south gradient in the prevalence of MS in both hemispheres,
the disease being rare in tropical regions and more common in
temperate climates. The gradient can be seen within a single
country. In Great Britain, for example, the prevalence of MS
in Cornwall (latitude 50°) is 63 per 100,000, while in Aberdeen
(latitude 57°) it is 144, and in the Orkney Islands (latitude
59°) it is 309 per 100,000 (15,18).

The fact that for patients of Northern European origin, and for
Jews, the risk of developing MS is strongly influenced by place
of residence in childhood, suggests that there is something in
the environment which is causally related to the disease. Its
nature has been the subject of many investigations and a large
number of statistically significant correlations have been
found between the prevalence of MS and both geo-climatic and
socioeconomic variables. The problem is to decide which factors are likely to be relevant to the disease. Liebowitz and
Alter (12) have argued convincingly that the epidemiological

studies in Israel are particularly relevant because conditions
there constitute an exception to the general rule that countries
of low latitude have a poor socioeconomic level of development.
In Israel, the development is high compared with that of the
surrounding countries. The prevalence of MS is also relatively
high (15 per 100,000), suggesting that the socioeconomic conditions are of greater importance than the shared geo-climatic
conditions. They further argue that the way in which the socioeconomic factors are most likely to operate is in relation to
exposure to infection.

It should be noted that there are exceptions to these general
trends. The rarity of MS in oriental countries will be discussed below. Clusters of unusually high prevalence have been
reported, but whether they are of purely statistical rather than
of biological significance is unknown (15).

The nature of the putative infective agent in MS is unknown.
The pathological characteristics of the disease are more in
keeping with infection with a virus than with any other known
type of agent. Several viruses having the properties which
would be required of an organism causing multiple sclerosis have
been isolated from animals (R.T. Johnson, personal communication),
but no single virus with all these characteristics have so far
been identified in man (11). Moreover, none of the lines of
evidence so far put forward implicating a virus in MS is conclusive: the elevated antibody titres to a number of viruses
are not specific to MS, comparable levels being found in chronic
liver disease and systemic lupus erythematosis; there has been
no consistent or confirmed isolation of virus from patients dying
with the disease; virus-like particles are not always present;
and intracerebral inoculation of material from patients with
MS into animals has provided no evidence for the presence of a
slow virus (10,11).

In summary, the evidence for the existence of an environmental
factor in the etiology of multiple sclerosis is strong. The
evidence as to its nature is weak. It is likely to be infective

and may be a virus, but we know of no organism which has all the necessary properties for producing an illness like MS. It is an open question whether there exists some kind of novel agent with properties not yet defined or whether there might be some mode of reaction of the body to infection with known agents that still eludes us.

Genetic Factors

Epidemiological studies also provide some evidence for the existence of a genetic factor. Multiple sclerosis is rare in the Japanese. The prevalence in Japan itself (4 per 100,000) is an order of magnitude less than in Europeans living at a comparable latitude (1). The prevalence in US-born Japanese is similar and is unrelated to latitude (7).

Further evidence for a genetic factor comes from family studies. Approximately 10% of the patients with multiple sclerosis have an affected close relative. Although twin studies present difficulties because of the possibility of bias introduced by the mode of ascertainment, three major studies show the same trend. Williams et al. (22) have summarized the position by stating that the concordance rate in clinical MS in monozygotic twins examined at any one time lies between 20 and 40% and in dizygotic twins is no greater than 15%.

This conclusion provides strong confirmation of the existence of a genetic factor in the etiology of MS.

Nature of the Genetic Factor

Present evidence suggests that it is related to the functioning of the sixth chromosome and in particular to that of the HLA system - i.e., in broad terms to the regulation of immune responsiveness (4). It is now known from the study of the analogous H2-system in the mouse that the gene products of this system are intimately involved at a number of stages in the formation of antibodies against certain antigens and in subsequent recognition and destruction of cells containing these antigens.

All the evidence suggests that the functions of the HLA system
in man are similar.

A significant association has been demonstrated between the HLA
system and MS. In Caucasoids there is a weak association with
HLA-A3 and B7 and a stronger association with the related antigens Dw2 and DR2. About 60% of multiple sclerosis patients have
the latter antigens compared with about 20% of the normal population (5). In the two Mediterranean groups examined, an association has been found with DR4 (5). There appear to be two
possible explanations for these specifically different associations. First, that neither gene is conferring susceptibility
to multiple sclerosis but that each is acting as a marker in the
appropriate population for an unidentified gene. Second, that
each gene is conferring susceptibility in the appropriate population but the environmenal agent is different. The evidence
from family and twin studies suggests that the former is the
more likely explanation. Among 197 patients in 91 families on
whom data has been published, approximately 20% of the individuals have different HLA profiles from their affected family
members (Compston, personal communication). Moreover, in the
study by Williams et al. (23), two pairs of dizygotic twins
concordant for MS did not share the same HLA antigens.

Pathogenesis
Any discussion of the mode of interaction between the genetic
and environmental factors must be highly speculative. In the
first place, the nature of the environmental factor is unknown.
Second, although the genetic factor is probably closely related
to the HLA system, it has not been identified. While many of
the closely related genes on the sixth chromosome will be related to the same functions - i.e., regulation of the immune
response - there is already evidence that others are not. Third,
although immunological abnormalities exist in MS, they are not
yet well characterized. CNS synthesis of oligoclonal gamma
globulin is almost universal, but the antigen has not been
identified.

Antel et al. (3) have recently demonstrated changes in lymphocyte sub-populations during relapses, but the relationship between these changes and the production of demyelination remains
to be determined.

The following is a guess as to what might be happening in MS
(4). Epidemiological evidence suggests that the crucial event
is exposure to an infective agent in childhood. The majority
of infected children would recover from the infection and develop lifelong immunity. A few would develop modified disease
leaving the individual sensitized to CNS components or with
persistent viral infection, or both. Later, in adult life,
a variety of events might evoke secondary immune responses
directed against brain components (an autoimmune mechanism
possibly analagous to relapsing EAE) or alternatively against
brain cells infected with reactivated virus, resulting in the
formation of a plaque. The HLA system could operate at two
levels in this scheme. A particular genotype might determine
whether or not an individual developed the modified disease,
and the HLA system might be involved in determining the sensitivity of the immunological memory to non-specific reactiva-
tion during adult life.

An unsatisfactory aspect of our present understanding of MS
is the rarity of the disease compared with the frequency in the
normal population of the genetic markers so far identified.
There are several possible explanations. First, if an unidentified single gene (for which Dw2, DR2, and DR4 are acting as
markers) is conferring susceptibility, it may be a rare one.
Second, the development of the disease may depend upon exposure
to a critical "dose" of an environmental agent. Third, there
may be additional genetic factors involved in the pathogenesis;
the epidemiological evidence for a protective factor among
orientals has already been mentioned and there are hints of
other susceptibility factors in Caucasoids.

SUMMARY
The evidence that the etiology of multiple sclerosis depends on
the interaction of environmental and genetic facts is good.

The nature of neither is known, but the former is probably infective, possibly viral, and the latter is probably related to
immune regulation. In both cases, the data is inadequate to
permit a decision as to whether one or more factors are involved in each class.

TREATMENT (8,13)
There are two aspects of the treatment of multiple sclerosis.
First, there is the relief of symptoms and of the social, economic, and psychological consequences of the disease, and the
treatment of complications.

The duration of individual episodes can be shortened by the
use of steroids but the functional outcome is unchanged. The
possibility of symptomatic treatment by the use of drugs which
help to restore conduction in blocked fibers is discussed in
a later paper.

The other aspect of treatment of multiple sclerosis is modification of the course of the disease. It is not surprising that
there is at present no established means of doing so. The
difficulties of evaluation of any treatment regime are obvious
from the variability of the course of the illness. Serious
attempts have nevertheless been made to assess the effects of
immunosuppression, selective immune stimulation (with transfer
factor), and dietary supplements of polyunsaturated fatty acids.
There is no convincing evidence yet that any of these approaches
is beneficial.

REFERENCES

(1)   Acheson, E.D.   1977.   Epidemiology of multiple sclerosis.
      Br. Med. Bull. 33: 9-14.

(2)* Adams, C.W.M.   1977.   Pathology of multiple sclerosis:
      progression of the lesion.   Br. Med. Bull. 33: 15-20.

(3)* Antel, J.P.; Richman, D.P.; Medof, W.E.; and Arnason, B.G.W.
      1978.   Lymphocyte function and the role of regulator cells
      in multiple sclerosis.   Neurology (Minn) 28(2): 106-110.

(4)* Batchelor, J.R.; Compston, A.; and McDonald, W.I.   1978.
      The significance of the association between HLA and multi-
      ple sclerosis.   Br. Med. Bull. 34: 279-284.

(5)   Compston, A.   1980.   Immunological mechanisms in neurologi-
      cal disease.   In Advanced Medicine, ed. A.J. Bellingham,
      vol. 16, pp. 314-323.   Tunbridge Wells: Pitman Medical.

(6)   Cuzner, M.L., and Davison, A.N.   1979.   The scientific
      basis of multiple sclerosis.   Molec. Aspects Med. 2: 147-248.

(7)   Detels, R.B.; Brody, J.A.; and Edgar, A.H.   1972.   Multiple
      sclerosis among American Japanese and Chinese migrants to
      California and Washington.   J. Chron. Dis. 25: 3-10.

(8)   Ellison, G.W., and Myers, L.W.   1978.   A review of systemic
      non-specific immunosuppressive treatment of multiple sclero-
      sis.   Neurology (Minn) 28(2): 132-139.

(9)   Esiri, M.   1977.   Immunoglobulin-containing cells in multi-
      ple sclerosis plaques.   Lancet 1: 478-480.

(10) Fraser, K.B.   1975.   Population serology.   In Multiple
      Sclerosis Research, eds. A.N. Davison, J.H. Humphrey,
      A.L. Liversedge, W.I. McDonald, and J.S. Porterfield, pp.
      53-67.   London: Her Majesty's Stationery Office.

(11)* Johnson, R.T.   1975.   Summary of virological data support-
      ing the viral hypothesis in multiple sclerosis.   In Multiple
      Sclerosis Research, ed. A.N. Davison, J.H. Humphrey, A.L.
      Liversedge, W.I. McDonald, and J.S. Porterfield, pp. 155-
      177.   London: Her Majesty's Stationery Office.

(12) Liebowitz, U., and Alter, M.   1973.   Multiple Sclerosis:
      clues to its cause.   Amsterdam: North-Holland.

(13) Liversedge, L.A.   1977.   Treatment and management of multi-
      ple sclerosis.   Br. Med. Bull. 33: 78-83.

(14) Lumsden, C.E.   1970.   The neuropathology of multiple sclero-
      sis.   In Handbook of Clinical Neurology: Multiple Sclerosis
      and Other Demyelinating Diseases, eds. P.J. Vinken and G.W.
      Bruyn, vol. 9, pp. 217-309.   Amsterdam: North-Holland.

(15)   McAlpine, D.; Lumsden, C.E.; and Acheson, E.D. 1972.
        Multiple Sclerosis: A Reappraisal, 2nd ed. London:
        Churchill Livingstone.

(16)*   McDonald, W.I., and Halliday, A.M. 1977. Diagnosis and
        classification of multiple sclerosis. Br. Med. Bull. 33:
        4-9.

(17)   Poser, S. 1978. Multiple Sclerosis: An Analysis of 812
        Cases, by Means of Electronic Data Processing. Berlin:
        Springer.

(18)   Poskanzer, D.C.; Prenny, L.B.; Sheridan, J.L.; and Kondy,
        J.Y. 1980. Multiple sclerosis in the Orkney and Shetland
        Islands. I: Epidemiology, clinical factors and methodology.
        J. Epidemiol. and Commun. Health 34: 229-239.

(19)   Prineas, J.W., and Connell, F. 1978. The fine structure
        of chronically active multiple sclerosis plaques. Neurology
        (Minn) 28(2): 69-75.

(20)   Prineas, J., and Connell, F. 1979. Remyelination in
        multiple sclerosis. Ann. Neurol. 5: 22-31.

(21)   Tanaka, R.; Iwasaki, Y.; and Koprowski, H. 1975. Ultra-
        structural studies of perivascular cuffing cells in mul-
        tiple sclerosis brain. Am. J. Pathol. 81: 467-474.

(22)   Thompson, E.J. 1977. Laboratory diagnosis of multiple
        sclerosis: immunological and biochemical aspects. Br.
        Med. Bull. 33: 28-33.

(23)   Williams, A.; Eldridge, R.; McFarland, H.; Houff, S.;
        Krebbs, H.; and McFarlin, D. 1980. An investigation
        of multiple sclerosis in twins. Neurology (Minn), in press.

* Short Reading List.

Neuronal-glial Cell Interrelationships, ed. T.A. Sears, pp. 25-38.
Dahlem Konferenzen 1982. Berlin, Heidelberg, New York: Springer-Verlag.

The Role of Neuronal-Glial Cell Interaction During Brain Development

P. Rakic
Yale School of Medicine, Section of Neuroanatomy
New Haven, CT 06510, USA

Abstract. Recent immunocytochemical studies using glial specific marker show that neuronal and glial cell classes coexist in
the developing primate brain earlier than has hitherto been assumed, and that they originate from separate precursors that are
present in the proliferative zones. DNA labeling indicates that
one transient subclass - the radial glial cells - does not divide
for about two months during the peak period of neurogenesis.
During this period their elongated processes may play a crucial
role in the compartmentalization of the nervous system and may
serve as guides for migrating neurons as they traverse the distance between their sites of origin near the ventricular surface
and their final destinations. The cellular mechanisms responsible
for this movement, as well as the possible involvement of glial
fibers in the formation of extracellular spaces and in the transfer of information and/or nutrients via intracytoplasmic transport during development, remain to be clarified. At somewhat
later stages the astrocytes that are produced directly from glial
precursors or by a morphogenetic transformation from radial glial
cells, construct the membrana limitans gliae that forms the brain
surface, separate differentiating neurons during synaptogenesis,
and remove degenerating cells, axons, and synaptic terminals.
Genetic or acquired abnomalities of glial cells may impair the
completion of the normal developmental process and lead to various brain abnormalities.

INTRODUCTION

It is now clear that from a very early stage the developing vertebrate brain contains at least two major cell classes - neurons
and glial cells. For many decades the prevailing view was that
glial cells are only formed after all (or most) of the neurons

destined for a given structure are generated. This view has now
been refuted by the use of sensitive glial specific markers.
Although it has become increasingly evident that glial cells
play an important role in neurogenesis from the earliest formative stages, many issues, including details of glial cell
lineage, the nature of their interaction with neurons, and their
function during various phases of ontogenetic development need
to be clarified. For example, when and where do the neuronal
and glial cell lines diverge? Are the glial and neuronal phenotypes expressed prior to or after the cessation of cell division?
What is the function of the transient radial glial cells? What
regulates the length of their dormant amitotic cycle? What
other functions do glial cells serve during normal and pathological development? In this background paper I shall discuss
some of these issues, mostly based on the work in my laboratory.
As a consequence I shall particularly emphasize the role of
neuron glial interaction during development of the primate brain.

WHEN AND WHERE DO GLIAL AND NEURONAL CELL LINES DIVERGE?
Since both neurons and the macroglial cells of the central nervous system are derived from the neuroepithelium of the primitive
neural tube, it is obvious that they must form separate cell
lines at some time during ontogenetic development. It has proven
to be rather difficult, however, to determine just when and
where this divergence occurs and this uncertainty has persisted
for almost a century. In his classical study, published in 1889,
His (9) proposed that the germinative epithelium of the developing cerebral wall consists of two classes of precursor cells,
one neuronal (his Keimzellen or "germinal cells") and the other
glial (his "spongioblasts"). This view is illustrated schema-
tically in Fig. 1A. An opposing view was suggested by Schaper
in 1897 (26). According to this the germinal zone consists of
a single precursor cell type (Fig. 1B) that, in time, gave rise
to both neurons and glia. This view seems to have prevailed
because His erroneously considered that the spongioblasts form
a multinuclear syncytium. The cellular homogeneity of the ventricular zone gained support from histological studies of

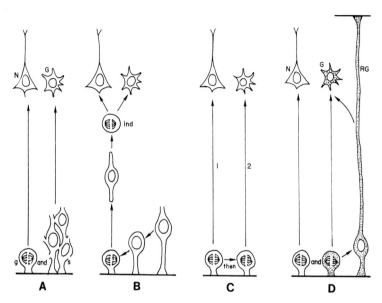

FIG. 1 - To illustrate previous theories of the origin of neuronal
and glial cell lines (A, B, C) and the scheme (D) suggested recently by Levitt, Cooper, and Rakic (12). A. Almost a century ago
His (9) distinguished two separate cell lines in the germinal matrix (or ventricular zone) lining the embryonic ventricles: the
round "germinal cells" (g), with mitotic nuclei which lie close
to the ventricular surface and give rise to neurons (N), and the
"spongioblasts" (S) whose nuclei lie at various distances from the
ventricle. The "spongioblasts," which His thought formed a syncytium, give rise to glia (G). B. A few years later Schaper (26)
proposed that these two "cell types" simply represented different
mitotic phases of a single cell class. According to Schaper, cells
of the ventricular zone produce indifferent cells (ind) that migrate into the intermediate (or mantle) zone where they further
divide into either neurons or both neurons and glial cells. C.
The results of tritiated thymidine autoradiography led Fujita (5)
to suggest that the dividing cells in the ventricular zone first
give rise to neurons (1); then after neuronogenesis has ceased,
the same dividing population begins to produce glial cells (2).
D. The localization of glial fibrillary acidic protein (GFA) by
electron microscopic immunohistochemistry (12) has demonstrated
that both GFA-positive (stippled) and GFA-negative mitotic cells
coexist in the ventricular zone during midgestation in the monkey
occipital lobe. The GFA positive cells first produce radial glial
cells (RG) and later either directly, or indirectly, astrocytes
or various specialized astrocyte-like cells.

Sauer (25) and later by electron microscopic (6) and ^{3}H-thymidine
(^{3}H-TdR) autoradiographic analyses (5, 29). In addition, since
studies of the time of cell origin showed few labeled glial cells
in adult animals that had been exposed to radioactive thymidine
as embryos, some investigators suggested that glial cells are
only generated after all, or most, of the neurons destined for a
given structure have been formed. Fujita (5, 6) proposed such a
rigid scheme for the successive generation of neurons and glia
as illustrated in Fig. 1C. It is now evident that routine light
microscopic observations and transmission electron microscopy
cannot reveal differences between glial and neuronal cell lines
at early developmental stages. Furthermore, the problem of the
dilution of label in the interpretation of ^{3}H-TdR studies makes
this method unsuitable for determining the time of emergence of
the first glial cells in a given region of the nervous system
(see discussion in refs. (13, 40)).

It should be pointed out that for some years there has been at
least one line of evidence that glial and neuronal cell lines
coexist during embryonic development. This evidence was provided
by the description of radial glial cells in the fetal brain.
This transient cell class was first described in Golgi preparations
towards the end of the last century (see review in (28)) and
was later shown electron microscopically during the middle and
late stages of primate corticogenesis (18). The bodies of the
radial glial cells are usually located near the ventricular
surface, and their elongated processes traverse the entire width
of the brain wall and terminate at the pial surface (Fig. 2).
Since neither the Golgi method, nor electron microscopy, can establish the nature or identity of such cells at early stages, we
have recently used an antibody to glial fibrillary acidic protein
(GFA) which has a high specificity for astroglial cells (4). Using
the peroxidase-antiperoxidase-immunocytochemical method of
Sternberger et al. (34), we have demonstrated the existence of radial
glial cells during the first third of gestation in all major subdivi-
sions of the embryonic primate brain (13). Because both glial cells
and neurons increase in number in the course of corticogenesis between embryonic days E40 and E100 (19), their separate precursor cells

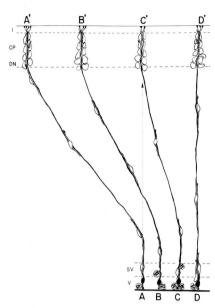

FIG. 2 - Schematic drawing of four radial glial cells and cohorts of associated migrating neurons. The cells located between the illustrated columns have been deleted in order to simplify the diagram and to emphasize the point that all neurons produced in the ventricular (V) and subventricular (SV) zones at the same site (proliferative units A-D) migrate in succession along the same fiber fascicle to the developing cortical plate (CP) and establish ontogenetic radial columns (A'-D'). Within each column the newly-arrived neurons bypass more deeply situated neurons (DN) that were generated earlier, and come to occupy the superficial-most position at the borderline between the developing cortical plate and the first cortical layer (1). Radial glial grids preserve the topographic relationships between generations of neurons produced in proliferative units at their final positions and prevent mismatching that could occur, for example, between unit A and Column C' if the cells were to take a direct, straight route (this indicated by the dotted line).

must co-exist. However, to identify glial precursor cells in the ventricular and subventricular zones, we had to use the immuno-cytochemical method at the electron microscopic level (12). This showed that the ventricular zone consists of two distinct classes of proliferative cells, and that at least in this respect, His' concept (9) of the coexistence of independent glial and neuronal cell lines in the germinal matrix seems to be correct (Fig. 1D). It should be pointed out that we have not been able to determine the exact time when the two cell lines diverge, since GFA may not be present at earlier stages even though the cells may already be determined. It is quite probable that the discovery of new glial or neuronal markers will reveal the divergence of the two cell lines at even earlier times. However, the fact that both cell classes are produced concomitantly, and in almost equal numbers, during the course of corticogenesis, makes it possible for us to address several additional questions as will be discussed in the following paragraphs.

IS THE GLIAL PHENOTYPE EXPRESSED BEFORE OR AFTER THE CESSATION
OF CELL DIVISION?

Light microscopic analyses, including immunocytochemistry, provide evidence that both neuronal and glial cell lines coexist
in the embryonic brain during the peak of neuronogenesis; however, they cannot resolve the issue whether the two cell classes
arise from common or from separate precursor cells. Until our
recent work (12,13) there was little firm evidence to challenge
Schaper's (26) notion that cell division in the proliferative zones produces "indifferent cells" which migrate across
the intermediate zone and only become transformed into neurons
or glia at their final destinations. To address this issue we
have examined the cytology of the ventricular zone with electron
microscopic immunocytochemical methods, again using an antibody
against GFA as a specific glial cell marker (12). This study
has shown that there are at least two mitotically active precursor cell types in the embryonic ventricular zone (i.e., GFA-
positive and GFA-negative cells). The localization of this
antigen in cells which are synthesizing DNA and actively dividing, indicates that proliferative cells are committed to their
specific lineages prior to the cessation of division. That is
to say the expression of these specific cellular phenotypes is
not critically dependent upon their final round of DNA synthesis. This is perhaps not unexpected since it is now known that
the differentiation of certain non-neural cell and sympathetic
neurons occurs before cell division (see (23, 24) for review).

Although at present we still do not know whether all or only
some GFA cells are transformed directly into astrocytes (Fig. 1D),
many such cells certainly pass through the transient stage of
radial glial cells. Using ^{3}H-TdR autoradiography, we have
shown that many radial glial cells remain in a dormant stage
for up to two months during midgestation in the rhesus monkey
(27), and that at some point they re-enter the mitotic cycle
and subsequently become transformed into astrocytes (20, 28).
Thus some of the mysteries of gliogenesis are being unraveled
although there are still many unresolved questions. For example,

we do not know whether the GFA negative cells consist only of neuronal cell lines or of multiple cell classes. At present, we cannot exclude the possibility that among the GFA negative cells are "indifferent" cells of the type envisaged by Schaper. Although we have established that neuronal and glial lines co- exist in the proliferative zones during the first half of gestation, the earlier lineage relationships between these cells will remain unclear until proliferative cells can be labeled by specific markers before the first neurons undergo their final cell division.

WHAT IS THE FUNCTION OF THE TRANSIENT RADIAL GLIAL CELLS?
The geometric regularity and the remarkable length of the radial fibers, which may be as long as 15-20 mm in the developing mon- key cerebrum, suggest that the functions of the radial glial cells may be related to some form of interaction with the cells that are situated along their trajectories. It was first sug- gested that the radial glial cells may be involved in transport processes (16); this idea is rather attractive since it suggests that the radial glial cells communicate information between the final positions that the neurons in any given structure will come to occupy and the proliferative zone in which the neurons are generated. This information could be carried by the trans- port of certain molecules as it has been recently shown that the radial glial cells have a considerable capacity for retro- grade transport from their endfeet to their cell bodies (3, 19). The formation of the glia limitans at the cerebral surface and at blood vessels (18, 28) may also provide some form of barrier between the cerebrospinal fluid-brain and blood-brain barriers, respectively (16). The trophic interactions between astrocytes and developing neurons that has been demonstrated in tissue culture studies (40) may also be extended to the radial glial cells. Immature glial cells may also influence the regulation of the extracellular fluid, as has been suggested for astrocytes in the adult brain (11). This may be especially important in developing nervous tissue where extracellular spaces may play a role in allowing growth, or even in directing, the elongation of nerve processes (30).

Other possible functions of the radial glial cells have been
discussed elsewhere (28). Here, I shall only emphasize their
role in guiding migrating neurons from their sites of origin
to their final positions in the cerebral and cerebellar cortices (17, 18). This concept is based on the ultrastructural
observation that migrating neurons remain attached to neighboring glial cells throughout their trajectories (see Fig. 2).
Such a relationship has been observed in the cerebellar and
cerebral cortex of many species (e.g., (7, 14, 21, 33)), as
well as in the hippocampal formation (15). The cellular relationships are more complex in the cerebellar cortex where
the granule cells pass through several morphogenetic transfor-
mations, but the general principle seems to be the same (17).
In addition to guiding neurons through the densely packed
neuropil, the radial glial fibers may ensure the faithful
mapping of the ventricular surface onto the expanding and con-
voluted primate cerebral cortex by preventing the lateral inter-
mixing of cells that are generated in different regions (20).
This model minimizes the amount of genetic information needed
for each neuron to reach its correct locus and to establish
its correct synaptic relations within the cortex.

WHAT IS THE NATURE OF THE GLIAL-MIGRATING CELL INTERACTION?
Transmission electron microscopy displays only the surface
apposition of the migrating neurons and radial glial fibers
that are separated by intermembranous spaces that are about 200-
300 Å wide (17, 18). No intercellular specializations are visible
by transmission electron microscopy, and to my knowledge freeze-
fracture experiments have not yet been done. It is noteworthy
that morphologically similar appositions exist between other
migrating cells and adjacent cell surfaces. For example, in
the monkey cerebrum we have counted as many as 1000 non-glial
cellular processes that are in direct contact. The fact that,
despite such numerous contacts, migrating neurons remain attached exclusively to glial fibers, even when they take a
tortuous course, suggests that there is a considerable affinity
between the membranes of these two cell classes. Such adhesivity

is especially evident in those areas of the tissue where large
extracellular spaces separate the neurons from other cells
and processes, but in which the migrating cells and radial
glial fibers remain closely apposed (18). This affinity may
be nonspecific, in the sense that all migrating neurons may
have the same affinity for any radial glial cell; nevertheless,
the orderly arrangement of the cells would still be preserved
since each group of neurons would become and remain, attached
to the neighboring radial glial fibers (Fig. 2). We have
evidence which suggests that some migrating neurons may transfer to a nearby radial glial cell but not to other classes of
cells of cellular processes (22, 28).

How can one reconcile the apparent contradiction that these
two cell types show a strong affinity for each other and at
the same time permit the movement of one along the other? One
among several possibilities that could account for this phenomenon is that the membranes of two cells could be fixed
for some time at one point along their interface, and the
migrating cell could still move if it were to add new membrane
components to its growing tip (Fig. 3). According to this
model, the leading process of the neuron would progressively
extend along the radial glial fiber, and the neuronal nucleus
would move to a new position within the perikaryal cytoplasm
(Fig. 3 B, C). The rate of movement of migrating neurons in
the primate telencephalon varies between 0.5 and 4µm per hour,
and this rate is compatible with a capacity for generation
and insertion of new membrane along the leading process. Several observations give credence to such a mechanism including
the finding of an increase in the surface area of migrating
cells (22) and the usual growth of neuronal processes at their
tips (1). We have found that the surface area of postmitotic
cells increases dramatically during the process of elongation
and continues to increase at a slower rate as the cells approach
the cortical plate (22). One may speculate that while new
membrane is being added to the leading process, degradation
may occur at the trailing process. The attractiveness of this
model is that it can be tested with the aid of recently developed immunocytochemical methods.

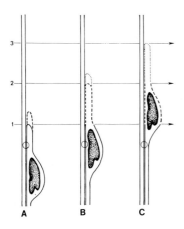

FIG. 3 - Diagram showing a possible
mechanism for the displacement of
migratory cells along the surface
of radial glial fibers (vertical
shafts A-C). To traverse the
distance between levels 1 and 2,
new membrane may be inserted a-
long the interface of the two
cells (dashed line in A), while
the nucleus moves within the
cytoplasm of the leading process
(B). As the leading tip grows to
reach level 3, additional new
membrane is inserted (dotted
line in C) along the glial surface and the nucleus moves to a
higher position between level
1 and 2, resulting in an overall
displacement of the cell body.
This model does not require movement of the neuronal surface along
glial fibers, and the binding sites
(circles) between the two apposing
membranes may remain constant for
sometime.

Although all the reported observations are suggestive of some form
of molecular binding between neurons and glial during migration,
such binding has not yet been directly demonstrated. If it does
indeed occur, we should like to know the biochemical and biophysical nature of the binding sites and the strengths of the
affinities between the cells in question. At present most developmental neurobiologists believe that most adhesions between
cells depend on some form of interaction between a component of
the cell surface and a complementary receptor on the apposing
cell surface; these components are widely thought to be either
glycoproteins or glycolipids (8, 35). The basic idea stems from
the work of Weiss (41) who suggested some years ago that cell
surfaces may contain complementary antigen-like and antibody-
like molecules that attract each other; unfortunately, unlike
certain other biological systems, it has not yet been possible
to isolate radial glial cells and migrating neurons and to study
their interactions in vitro. The use of antibodies that interfere with cell adhesion may be useful in this analysis (see (8)),
but such antibodies have not yet been generated.

A somewhat different approach to the molecular basis of this phenomenon would be to analyze neurogenesis in developing animals in which neuronal migration is impaired either by external agents or as the result of a single gene mutation (2, 21). For example, the autosomal recessive mutation in the homozygous weaver mouse (wv) leads to the defective migration and eventual death of most cerebellar granule cells (21). In this mutant, the Bergmann glial fibers (which are the cerebellar equivalents of radial glial fibers) display an ultrastructural abnormality during a circumscribed developmental period (21, 32). It is at present not clear whether the glial abnormality causes, or is caused by, the granule cell deficit (2), but on several grounds we have postulated that the glial cell is closer to the primary action of the defective gene (21). Recently Sommer et al. (31), using a monoclonal antibody (C1) against an antigen which is normally expressed in Bergmann and radial glial fibers, have found that this antigen is either not present, or is present in only minimal amounts, in several mutant mice, including 2-4 week old weaver animals. Since this glial antigen is usually expressed very early in development, it will be important to determine whether its depression occurs before or after the granule cells begin their migration and exhibit signs of degeneration, to clarify whether or not the glial defect is primary. The discovery of new antigens, both cytoplasmic or surface, that are unique to particular classes of cells may open new possibilities for further studying of glial and neuronal interactions in vitro and in vivo. At present we have evidence that developing neurons are highly dependent for their migration, growth, and maintenance on the glial cells in their immediate environment, but the mechanisms underlying this dependence remain to be determined.

REFERENCES

(1) Bray, D. 1973. Model for membrane movements in the growth
     cone. Nature 244: 93-96.

(2) Caviness, V.S., Jr., and Rakic, P. 1978. Mechanisms of
     cortical development: a view from mutations in mice. Ann.
     Rev. Neurosci. 1: 297-326.

(3) Chu-Wang, I-W.; Oppenheim, R.W.; and Furel, P. 1980.
     Ultrastructure of migrating spinal motoneurons in anurian
     larvae. Brain Res., in press.

(4) Eng, L.F., and Rubenstein, L.J. 1978. Contribution of
     immunohistochemistry to diagnostic problems of human
     cerebral tumors. J. Histochem. Cytochem. 26: 513-522.

(5) Fujita, D. 1963. The matrix cell and cytogenesis in the
     developing central nervous system. J. Comp. Neur. 120:
     37-42.

(6) Fujita, S. 1966. Application of light and electron micro-
     scopic autoradiography to the study of cytogenesis of the
     forebrain. In Evolution of the Forebrain. Phylogenesis and
     Ontogenesis of the Forebrain, eds. R. Hassler and H. Stephen,
     pp. 180-196, Stuttgart: Thieme.

(7) Gona, A.G. 1978. Ultrastructural studies on cerebellar
     histogenesis in the frog: The external granular layer
     and the molecular layer. Brain Res. 153: 435-447.

(8) Gottlieb, D.E., and Glaser, L. 1980. Cellular recognition
     during neural development. Ann. Rev. Neurosc. 3: 303-318.

(9) His, W. 1889. Die Neurobasten und deren Entstehung im
     embryonalen Mark. Abh. Math. Phys. Cl. Kgl. Sach. Ges.
     Wiss. 15: 313-372.

(10) Ivy, G.W., and Killackey, H.P. 1978. Transient population
     of glial cells in developing rat telencephalon revealed
     by horseradish peroxidase. Brain Res. 158: 213-218.

(11) Kuffler, S.W., and Nicholls, J.G. 1966. The physiology
     of neuroglial cells. Ergeb. Physiol. 57: 1-90.

(12) Levitt, P.; Cooper, M.L.; and Rakic, P. 1981. Coexistence
     of neuronal and glial precursor cells in the cerebral
     ventricular zone of the fetal monkey: An ultrastructural
     immunoperoxidase analysis. J. Neurosc., in press.

(13) Levitt, P.R., and Rakic, P. 1980. Immunoperoxidase
     localization of glial fibrillary acid protein in the
     embryonic rhesus monkey. J. Comp. Neur. 193: 815-840.

(14) Mugnaini, E., and Forstrnen, P.F. 1967. Ultrastructural studies on the differentiation of granule cells and development of Glomeruli in the chick embryo. Z. Zellforsch. 77: 115-143.

(15) Nowakowski, R.S., and Rakic, P. 1979. The mode of migration of neurons to the hippocampus: A Golgi and electron microscopic analysis in fetal rhesus monkey. J. Neurocytol. 8: 697-718.

(16) Oksche, A. 1968. Die pranatale und vergleichende Entwicklungsgeschichte der Neuroglia. Acta Neuropathologica, Suppl. 4: 4-19.

(17) Rakic, P. 1971. Neuron-glia relationship during granule cell migration in developing cerebellar cortex. A Golgi and electronmicroscopic study in Macacus rhesus. J. Comp. Neur. 141: 283-312.

(18) Rakic, P. 1972. Mode of cell migration to the superficial layers of fetal monkey neocortex. J. Comp. Neur. 145: 61-84.

(19) Rakic, P. 1974. Neurons in rhesus monkey visual cortex: systematic relation between time of origin and eventual disposition. Science 183: 425-427.

(20) Rakic, P. 1978. Neuronal migration and contact guidance in primate telencephalon. Postgrad. Med. J. 54: 25-40.

(21) Rakic, P., and Sidman, R.L. 1973. Sequence of developmental abnormalities leading to granule cell deficit in cerebellar cortex of weaver mutant mice. J. Comp. Neur. 152: 103-132.

(22) Rakic, P.; Stensaas, L.J.; Sayre, E.P.; and Sidman, R.L. 1974. Computer aided three-dimensional reconstruction and quantitative analysis of cells from serial electron microscopic montages of fetal monkey brain. Nature 250: 31-34.

(23) Rothman, T.P.; Gershon, M.D.; and Holtzer, H. 1978. The relationship of cell division to the acquisition of adrenergic characteristics by developing sympathetic ganglion cell precursors. Dev. Biol. 65: 322-341.

(24) Rutter, W.J.; Pictet, R.L.; and Morris, P.W. 1973. Toward molecular mechanisms of developmental processes. Ann. Rev. Biochem. 42: 601-646.

(25) Sauer, F.C. 1935. Mitosis in the Neural Tube. J. Comp. Neur. 62: 377-405.

(26) Schaper, A. 1897. Die frühesten Differenzierungsvorgänge im Central nervensystem. Arch. Entwickl.-Mech. Orig. 5: 81-132.

(27) Schmechel, D.E., and Rakic, P. 1979. Arrested proliferation of radial glial cells during midgestation in rhesus
monkey. Nature (London) 277: 303-305.

(28) Schmechel, D.E., and Rakic, P. 1979. A Golgi study of
radial glial cells in developing monkey telencephalon:
Morphogenesis and transformation into astrocytes. Anat.
Embryol. 156: 115-152.

(29) Sidman, R.L.; Maile, I.L.; and Feder, N. 1959. Cell proliferation in the primitive ependymal zone; an autoradiographic study of histogenesis in the nervous system. Exp.
Neurol. 1: 322-333.

(30) Singer, M; Nordlander, R.H.; and Egar, M. 1979. Axonal
guidance embryogenesis and regeneration in the spinal
cord of the newt: The blueprint hypothesis of neural pathway patterning. J. Comp. Neur. 185: 1-22.

(31) Sommer, I.; Lovenauer, C; and Schachner, M. 1981. Selective recognition of Bergmann glial and ependymal cells in
the mouse nervous system by monoclonal antibody.

(32) Sotelo, C., and Changeux, J.P. 1974. Bergmann fibers and
granular cell migration in the cerebellum of homozygous
weaver mutant mouse. Brain Res. 77: 484-491.

(33) Stensaas, L.J. 1972. An electronmicroscopic study of the
organization of the cerebral cortex of the 60mm rabbit
embryo. Z. Anat. Entwickl.-Gesch. 137: 335.

(34) Sternberger, L.A.; Hardy, P.H.; Cuculis, J.J.; and Meyer,
H.G. 1970. The unlabeled antibody enzyme method of immunohistochemistry. Preparation and properties of soluble
antigen-antibody complex (horseradish peroxidase-anti-
horseradish peroxidase) and its use in identification of
spirochetes. J. Histochem. Cytochem. 18: 315-383.

(35) Subteberg, S., and Wessels, N.K., eds. 1980. The Cell
Surface: Mediator of Developmental Processes. New York:
Academic Press.

(36) Varon, S.S., and Somjen, G.G. 1979. Neuron-glial interaction. Neurosc. Res. Prog. Bull. 17: 47-65.

(37) Weiss, P. 1947. The problems of specificity in growth
and development. Yale J. Biol. Med. 19: 235-278.

Neuronal-glial Cell Interrelationships, ed. T.A. Sears, pp. 39-56.
Dahlem Konferenzen 1982. Berlin, Heidelberg, New York: Springer-Verlag.

Membrane Specializations in Neuroglial Cells and at Neuron-Glia Contacts

E. Mugnaini
Laboratory of Neuromorphology, Dept. of Biobehavioral Sciences
Box U-154, University of Connecticut, Storrs, CT 06268, USA

Abstract. The architectural and junctional specializations of
astrocytes, oligodendrocytes, and Schwann cells are reviewed
in the context of the general features of epithelia and tissue
compartments. Glial cells form microvilli, secretory vesicles,
and several types of cell junctions. Also, certain non-junctional membrane specializations have been revealed by the
freeze-fracture method. Most of the membrane specializations
are not randomly distributed, but appear to be related to local
peculiarities in the cell's microenvironment. The most conspicuous of these are found at Ranvier nodes of myelinated
fibers and at the interface between different tissue compartments. The individual specializations are briefly reviewed.
Possible relations to certain pathological processes that may
affect astrocytic orthogonal assemblies, gap junctions, tight
junctions, and nodal and paranodal specializations are stressed.

GENERAL CONSIDERATIONS

One of the primary goals of research on neuroglia is to elucidate
their normal relationships and the various alterations that
occur in these relationships which are found in a number of
neurological disorders.

The classification of cell types in a nonhomogeneous tissue
such as the CNS is fundamental and requires an understanding

of both their structural and functional properties. Cajal
stressed the distinction between what he termed "autonomous"
glial cells and "epithelial" glial cells. The former include
those cells which in the process of development lose their
relation to the lumen of the neural tube; the thicker the wall
of the neural tube or its later derivatives, the brain and the
spinal cord, the more complex the glial cells appear. This is
true both in lower and higher vertebrates. In certain relatively thin portions of the neural wall, there is often a kind
of glial cell, the tanycyte, which, in its morphology, is similar to the "primitive" ependymal cell, and in others typical
ependymo-glial cells persist that span the entire thickness of
the wall of the brain or spinal cord from the ventricular wall
to the pial surface. In the paper by M. Schachner et al. (this
volume), the existence of cell markers specific for certain
autonomous glial cells and epithelial glial cells is discussed.

Since it is with the various forms of autonomous glia that we
are primarily concerned, it is worthwhile stating at the out-
set that these cells are morphologically and perhaps also functionally, very complicated and difficult to study (15). It is
important to keep in mind that they are derived from an epithelium and may retain many of the general properties of epithelial
tissues. Some of these properties are conserved in the process
of differentiation, while others are clearly modified. The
primary reason for their modification is that they may be in-
compatible with the highly specialized state of the associated
neurons, and their relationship, in turn, may impose specific
structural specializations upon the glial cells. Similar re-
straints may be imposed upon them by their requirement to maintain homeostasis within the central or the peripheral nervous
system. Among the general properties of epithelia that may be
relevant are mechanical adhesion between the cellular elements,
the establishment of transepithelial ionic and metabolic gradients, cooperation between individual cells, and various secretory or absorptive functions. In simple stratified epithelia,
these properties determine the appearance of numerous membrane

specializations so that the plasma membranes of the cells become distinctly nonhomogeneous: for example, the luminal surface may have microvilli and cilia; the lateral surface tight
junctions, gap junctions, adherens junctions, and plasmalemmal
interdigitations; while the basal surface may be specialized
for certain absorptive or secretory functions. The "glycocalyx"
surrounding the cell is also anisotropically developed, and
certain intracellular organelles may also be polarized. Many
of these features are readily recognized in such epithelial
glial cells as embryonic radial glia, Bergman glia, and Müller
cells, but some of them are also apparent in autonomous glial
cells.

One special parameter that has to be considered for the non-
neuronal elements in the CNS is the degree of their contact
with neurons. This clearly varies in the different subclasses
of glial cells. The specialized ependymal cells that line the
choroid plexus represent one end of this spectrum in that they
have minimal contact with neurons (although it should be mentioned that they may be regulated by the recently discovered
neurosecretory axons); the myelin forming glia and the protoplasmic astrocytes are at the opposite end of the spectrum ex-
hibiting maximal contact. With few exceptions, the non-neuronal
elements also provide for the separation of the neurons from
the mesodermal elements within and around the CNS. This role
is evident at a very early stage in ontogeny, and the ensheathment of such mesodermal components as blood vessels and the
leptomeninges becomes quickly established.

The multiplicity and the complexities of cell shapes make the
analysis of the nonhomogeneity of glial cell plasma membranes
extremely difficult. However, some general features can now
be recognized. For example, astrocytic glial membranes that
are in contact with the pia mater or with blood vessels can
be seen in freeze-fracture replicas to contain large numbers
of special aggregates of intramembrane particles known as
orthogonal assemblies. Similarly, the outer surfaces of

Schwann cells present numerous caveolae, while their nodal
surfaces have microvilli; and at the marginal cytoplasmic belts
of the myelin sheaths, there are clear tight junctions (19).

Most of the specialized junctions occur between macroglial cells
of like kind. These junctions are distributed in a fairly reg-
ular manner in epithelial glia. On the other hand, certain cell
junctions such as the gap junctions appear to be quasi-random
in their distribution in the autonomous glia. Reflexive cell
junctions (i.e., junctions between two portions of the plasmalemma of the same cell) are especially numerous in the myelin
sheath, most often at specific sites. However, specialized
cell junctions also occur between glial cells and neurons, particularly at the paranodes and at neuronal cell bodies in corre-
spondence with the so-called subsurface cisterns. In the absence
of specific membrane probes, most other glial neuronal contacts
are difficult to characterize. For instance, although astrocytic
processes envelop synaptic profiles and seem to be attracted to
them at early stages of neural development, in freeze-fracture
replicas they do not show conspicuous specializations.

Similarly, it has been reported that during development, certain
migrating neurons tend to be oriented along radial glial fibers
(14), and this has led to the theory that the radial glial cells
are essential for determining the positional specificity of the
neurons (see Rakic, this volume). However, migrating neurons
and their neurites form fairly sparse adhering junctions with
the associated glia (8). At present our knowledge of membrane
glycoproteins and glycolipids and of the organization of the
glycocalyx is too rudimentary to explain their possible roles
in glial-neuronal relations (16, 22, 23).   It is unknown whether resting microglial cells (a category of glial cells that
is believed by some authors to be derived from the mesenchyme)
form specialized cell contacts with each other and with other
glial cells, or have special non-junctional plasma membrane features. Microglial cells have never been investigated by freeze-
fracturing.

VARIETIES OF CELL JUNCTIONS

In vertebrate neural tissues, various types of cell junctions
can be seen.   These are adherens junctions (desmosome, hemi-
desmosome, septate junction, intermediate junction), gap junctions, tight junctions, and special junctions such as the paranodal axo-glial junction, the glial-neuronal junction related
to subsurface cisterns, the synapse, and the adhesion between
myelin lamellae.

Macular desmosomes occur between ependymal cells, tanycytes, and
also astroglial cells in cold-blooded animals. Bundles of glial
intermediate filaments and microfilaments attach themselves to
the cytoplasmic plaques of these junctions. Astroglial desmosomes are usually absent in mammals where the glial intermediate
filaments do not show obvious relations to the cell limiting
membrane in standard thin sections. Perhaps in the intricate
neuropil of higher vertebrates other forms of non-junctional
intercellular adhesion between astrocytic elements are developed.

Zonular desmosomes are frequently found within the peripheral
myelin sheath, where they are of the reflexive type and bind
together cytoplasm-containing loops of the myelin-forming
lamella.

Hemidesmosomes are present at points along the membrana limitans
gliosa at the interface with the pia mater in all vertebrates.

Typical septate junctions in the mammalian nervous system have
been observed only between certain cerebellar axons.

Intermediate junctions often join adjacent neuronal cell processes and are usually referred to as puncta adherentia (15).
They also occur between like cells of the epithelial glial, the
astrocytic, the oligodendrocytic, and the ganglionic satellite
categories, and also between astrocytes and oligodendrocytes.
Rarely, intermediate junctions occur between the inner turn
of the myelin sheath and the axolemma. Assuming that the

paranodal axo-glial junction is not a special form of intermediate junction, adherens junctions between autonomous glial
cells and between glial cells and neurons are less conspicuous
than among the neuronal cell processes themselves. A special
feature should be remembered; namely, that the intermediate
junctions between cells of the astrocytic and oligodendrocytic
categories are often continuous with gap junctions. This feature has been observed in different kinds of epithelia. Also,
by virtue of their relations with the cytoskeleton, the adherens
junctions may be involved in the maintenance of cell shape and
other related functions.

Gap junctions. It has been known for a long time that gap junctions occur frequently, not only between ependymal cells, tanycytes, cells that form the epithelial lining of choroid plexus,
and radial glia cells, but also between the plasma membranes of
adjoining astrocytic cell bodies and processes. Such junctions
could mediate the redistribution of ions and small solutes,
such as certain neuron transmitters, from regions of high concentration to regions of low concentration, thus preventing accumulation of these substances around active neurons to levels that
would be detrimental to specific neuronal circuits. We have
recently observed that typical gap junctions also occur frequently between oligodendrocytes and astrocytes. These have
been seen both in thin sections (6) and freeze-fracture replicas
(Massa and Mugnaini, unpublished observations). Thus, astrocytes
and oligodendrocytes form a functional syncytium providing ionic
and metabolic coupling. It is therefore possible that disease
processes exist, which may affect myelin indirectly via the
astroglia. Specialized membrane appositions between interfascicular oligodendrocytes have been described in thin sections.
These appositions resemble gap junctions. The existence of gap
junctions between oligodendrocytes, however, has not been ascer-
tained in freeze-fracture replicas, and the nature of the special
appositions remains unclear.

Gap junctions have also been observed between an adjoining
myelin sheath and an astrocytic process, and within individual
myelin sheaths. The latter are of the reflexive type and occur
most commonly between paranodal loops. Unpublished studies in
the mouse and the cat nervous system suggest that reflexive
myelinic gap junctions may be limited to poikilotherms. Do these
junctions simply form as a corollary to adherens junctions,
which are also common between loops, or do they play a specific
role in the movement of ions entering the myelin sheath in relation to saltatory conduction? The questions whether astrocytic
and oligodendrocytic plasma membranes are equally permeable to
the same ions, and whether selective permeability exists on
different portions of the same cell remain to be answered
(Orkand, this volume).

Tight junctions. We have never observed tight junctions between
astrocytes, but in freeze-fracture replicas we have seen that
the plasma membranes of oligodendrocytic cell bodies can form
tight junctions with neighboring myelinated axons. Furthermore,
both in the CNS and PNS, tight junctions are regularly present
in myelin, at the borders of the sheath, along cytoplasmic
incisures, and, in central sheaths, between compact lamallae.
Usually tight junctions of the myelin sheath are reflexive and
are organized as non-anastomosed linear strands. Experiments
using electron dense tracers have shown that they are effective
barriers to the diffusion of large molecules but not for small
solutes. Three possible functions could be attributed to the
myelinic tight junctions: (a) they could form a seal preventing
equilibration of the intramyelinic extracellular space and the
extracellular compartment of the nervous system; (b) they may
hold together adjacent turns of the myelin lamella; or (c) they
may provide a barrier to lateral diffusion within the membrane
itself facilitating spatial nonhomogeneity.

In freeze-fracture replicas from mature animals, features
suggestive of the assembly or disassembly of tight junctions
have been observed. Recent evidence from thin section and

freeze-fracture studies indicates that tight junctions begin to
form, both in CNS and PNS, at the earliest stages of ensheathment
when primitive mesaxons first become recognizable (19). Tight
junctions could therefore be essential for both the formation
and the maintenance of the myelin sheath. Interestingly, tight
junctions at the inner mesaxon are not present until the sheath
has acquired several compact turns (unpublished observations).

The tight junctional protein, which is thought to be an integral
membrane protein, has not yet been identified. It is fair to
assume that one of the minor protein bands in central myelin
represents the tight junctional protein; monoclonal antibodies
may provide the means for its identification. All the myelin
deficient mice mutants so far explored have tight junctions within their myelin sheaths.

Tight junctions of the anastomotic variety (a variety often
correlated with an efficient barrier to extracellular diffusion)
also occur at the cell lining of choroid plexus, the tanycytes
of the median eminence and of the subcommissural organ, arach-
noideal and perineurial mesothelia, and the endothelia of
most neural blood vessels (12). Tight junctions are usually
absent between ependymal cells and radial glial cells. Since
tight junctions occur between distinct types of cells only, they
may be considered as a special kind of cell marker.

SPECIALIZED JUNCTIONS AND MEMBRANES IN NEURAL TISSUE
Neuron-glia Junctions at the Subsurface Cisterns
All nerve cells contain a varying number of special hypolemmal
endoplasmic cisterns, termed subsurface cisterns, that have
extremely narrow lumina and are closely apposed to the plasma
membrane (5). The great majority of subsurface cisterns are
observed facing a glial cell profile. The usual glial partner
of the subsurface cisterns in the CNS is the astrocyte and in
the PNS, the satellite cell. In freeze-fracture replicas, the
membranes of the subsurface cistern, the neuronal plasmalemma,
and the adjoining glial membrane at the site of mutual apposition

show a patch of intramembrane particles, suggesting the presence of a specialized cell junction. This junction may be
the site of Ca^+ release from the endoplasmic reticulum (7).

Neuron-glia Synapses
Specialized neuron-glia contacts may show clusters of synaptic-
like vesicles in the neuronal element that are obviously related to the cell junction. These neuron-glia "synapses" occur
rarely in mature animals, but may be common at certain stages
of development. Their function is unknown and is presumably
transitory. Perhaps their presence is related to cell surface
markers which are expressed only at early stages of development.
The presumed neuro-glial synapses involve the radial glia and
perhaps the astrocytes (8).

A special form of junctional apposition between glial and neuronal membranes that lacks synaptic vesicles is frequently seen in
mature nervous tissue from neurological mutants, from animals
recovering from neurological lesions, and in developing nervous
tissue. The neuronal elements in the close appositions resemble
normal postsynaptic specializations. The prevalent interpretation of these junctions is that they are "vacated" postsynaptic
sites; i.e., junctional postsynaptic sites remaining after degeneration or retraction of a presynaptic specialization that
become occupied by astrocytic or oligodendrocytic cell processes (20).

The Paranodal Axo-glial Junction
The internodal, paranodal, and nodal regions of many kinds of
myelinated fibers have been studied in considerable detail in
both CNS and PNS of cold-blooded and warm-blooded animals. The
basic principles of membrane organization in myelinated fibers
seem to be similar in CNS and in PNS and across the species,
but certain proven differences exist and some structural details
are a matter of dispute (17, 19, 21).

The internodal axonal and glial membranes usually lack conspicuous specializations. At the paranode, however, all
myelinated axons show a special cell junction that prevents
the passage of extracellular tracers of large molecular weight
from the nodal gap to the adaxonal space of the internodal
region of the fiber. The transition between internodal and
paranodal regions is quite abrupt in cold-blooded animals; in
homeotherms, one sees in freeze-fracture preparations special
clusters of intramembrane particles both in the axolemma and
in the glial membranes (unpublished observations).

Since experimental studies suggest that membrane turnover
and the exchange of phospholipids from the axon to the myelin
sheath (2) may take place preferentially in this region of
the fiber, the precise features of intramembrane particles in
the transitional zones should be resolved in greater detail.

The basic features of the paranodal axo-glial junctions are:
a) the axon at the paranode is considerably constricted (to
about 1/3 of the internodal diameter); b) the lateral cytoplasmic belt of the myelin sheath (which in sections forms the
paranodal loops) indents the axonal perimeter; c) the apposed
glial and axolemmal membranes form an unusually extensive cell
junction characterized by a narrow intercellular cleft (30Å)
and the presence of structured dense material in the cleft that
is organized in a repeat pattern; d) the junctional plasma membranes contain rows of intramembrane particles. The particles
display different cleaving properties in the glial and the
axonal membranes, indicating that the junction is structurally
asymmetric; e) the rows of junctional particles are oriented
parallel to the paranodal belt in warm-blooded animals, but
are disposed at an angle to the belt in poikilotherms. In
mammals, very small particles have been observed in the junctional glial membrane in the space between the rows of major
particles (2). These "subparticles" have not been seen in
cold-blooded animals; f) while some authors, who have studied
the distribution of extracellular tracers in thin sections,

assume that the intercellular cleft contains septa, others
maintain that the intramembrane particles themselves protrude
into the space and are responsible for the electron densities
seen in thin sections. If present, the septa would be aligned
with the rows of particles. Oblique orientation of the septa-
like structures in the intercellular space in poikylotherms
suggests that paranodal junctions may be more leaky in such
animals as the frog than in birds and mammals; g) in thicker
axons, the paranodal loops are narrow and many of them fail to
attach on the axolemma, while in thinner axons (up to 2.5 μm in
diameter), the loops are larger, and most of them are junction-
ally joined to the axolemma (Massa and Mugnaini, unpublished
observations). One may speculate therefore that larger fibers
are especially subject to paranodal demyelination. The acute
effects of paranodal demyelination could be particularly interesting to study with the freeze-fracture method, in combination
with electropharmacological procedures. In tissue culture preparations, the disruption of the paranodal axo-glial junction
by low-calcium media or by trypsin treatment produces retraction of the ends of the myelin sheaths towards the center of
the internode (see literature (19)).

The Nodes of Ranvier
The nodal and paranodal regions differ to some extent in the
CNS and PNS, although they maintain some well-known basic
similarities discussed below.

In thin sections, the nodal axolemma of central and peripheral
axons presents an electron dense microfibrillary undercoating
similar to that seen beneath the axolemma of the initial axon
segment. The possibilities that this represents cytoskeletal
elements anchoring ion channels in the axolemma remain to be
explored. The nodal axolemma has a high density of large
(100-250Å) intramembrane particles that cleave with both fracture faces (in the non-excitable nodes of electrocyte axons, as
in Sternarchus, few particles cleave with the E face).

The size, shape, and density of distribution of the intramembrane particles in the nodal axolemma differ little in
poikylotherms and homeotherms and range between 2500 and
$3000/\mu m^2$. Nevertheless, since most of the particles in the
internodal axolemma are small (80Å) and cleave with the protoplasmic face, it seems obvious that the nodal membrane in
excitable nodes is specialized. Direct evidence that E face
particles correspond to sodium channels is lacking. It is
difficult to be specific about the value of particle numbers
because not all of the intramembrane proteins may be revealed
by the standard freeze-fracture technique. Particle shape and
size and cleaving properties are also most often difficult to
correlate with precise biophysical data. The new technique
developed by Heuser and Reese, which involves rotary shadowing
of rapidly frozen, deeply etched cells or tissue, has revealed
that the concentration of acetylcholine receptors (9) at a
motor and plate is on the order of $10,000/\mu m^2$; the application
of this technique may provide useful data for the Na^+ channels.

In the PNS, there is much more nodal gap substance than in the
CNS. This is made up of charged components with different ion binding properties (11). In the CNS, an astrocytic process (that is
part of the glial syncytium) often approaches the nodal axolemma.
In the PNS, on the other hand, the last turn of the Schwann cell
sheath gives rise to digitiform processes in smaller fibers, and
to true microvilli in larger fibers. The larger the axonal diameter, the higher the number of microvilli. In the large peripheral fibers one finds, at the outer myelin turn and near the
node, distinct cytoplasmic pockets that are filled with mitochondria; this appearance suggests that energy (presumable for
ionic pumps) is provided at these sites. Such pockets are
absent in the CNS.

In both the CNS and PNS, the paranodal loops contain tubules
of endoplasmic reticulum that may be associated with the axo-
glial junctions. While in many fibers this reticulum is formed
by short tubules, in some peripheral nerve fibers there is a

long tubule that may span the entire length of the paranodal
belt. The function of these glial paranodal endoplasmic tubules
is unknown. They may participate in ion transfer. Ellisman and
his co-workers (4) have studied myelinated fibers in the PNS by
means of cytochemical cation precipitation techniques and electron
probe wavelength spectroscopy. They conclude that the paranodal
loops contain elevated concentrations of Na^+ and Ca^+, and that
elevated levels of Ca^+ are also present in compact myelin. They
have proposed a Na^+ pathway involving the nodal axolemma, the
nodal villi, the paranodal cytoplasm, and the axo-glial junctional
membranes. This tentative model serves to emphasize that much
remains to be learned about the neuroglial signals that are transferred at different regions along the length of myelinated fibers.

The ways in which Na^+ channels are inserted into the axonal membrane, how they become concentrated at the nodes of Ranvier, and
their rate of turn-over are unknown. Bray and Oldfield (personal
communication), using freeze-fracturing, have observed that the
number of axolemmal particles shows a marked increase in the
axons of the rat optic nerve between postnatal days 7 and 14.
Accumulations of particles resembling those seen at nodes of
Ranvier were not observed along unensheathed axons.

The distribution and the morphology of axolemmal particles have
also been studied with freeze-fracturing in murine dystrophic
mutants with defects in myelinogenesis. Rosenbluth (18) has
concluded that patches of nodal-like axolemmal particles occur
only where axons are in intimate contact with myelinating cells.
Ellisman (3), however, is of the opinion that the occurence of
a nodal-like particle profiles on the axolemma of dystrophic mice
is not dependent upon the presence or absence of Schwann cells.

Thus, we still do not know what differences there are (apart from
size) between myelinated fibers and fibers that remain unmyelinated, nor do we know what determines ensheathment. The possible
involvement of the glial and axonal cytoskeleton in the processes
of ensheathment and myelin formation remains to be analyzed (see
Bunge, this volume).

The Compact Myelin Membrane

Membrane adhesion in myelin assumes unusual forms (13). Large proportions of the cytoplasmic surfaces of myelin lamellae are joined together to form the major dense lines. The exoplasmic surfaces of neighboring lammellar turns are also closely apposed, much more so than in most specialized cell junctions with the exception of the zonulae occludentes. The myelinic intercellular gap is thinner in central than in peripheral sheaths. Intramembrane particles abound both in central and peripheral myelin, but their size distribution, shape, and cleaving properties differ in the two situations. Central myelin particles cleave preferentially with the P face and occur mainly in two forms, round and elongated; in peripheral myelin (which contains much more glycoprotein than central myelin), numerous particles cleave with the E face.

CONCLUSIONS

Within the nervous system, non-neuronal cells of various types, by virtue of cell junctions and other specialized membrane features, appear to establish a complex pattern of compartments which may have special roles during development, in the maintenance of the tissue, and after its injury.

Various types of glial cells in the CNS form gap junctions with cells of like categories, and the astrocytes often form gap junctions among themselves and with oligodendrocytes. The gap junctions establish intracellular channels of communication through which ions and small solutes (below 1300 Daltons) can pass. On the other hand, in peripheral nerves of mature normal animals, individual Schwann cells do not come into direct contact with each other.

In the PNS the myelin forming cells abut directly onto connective tissue surfaces. However, in the CNS only the astrocytes and the epithelial glial cells are in direct contact with leptomeninges, or the walls of blood vessels. Astrocytes at the interface with connective tissue elements do not establish barriers to paracellular diffusion. But the astrocytic membrane at the interface with the connective tissue does appear to be specialized; this specialization has been revealed by freeze-fracturing and consists of the accumulation of special orthogonal assemblies of small intramembranous particles that, on occasion, have a paracrystalline configuration. Similar orthogonal assemblies have also been found in a number of other epithelia as well as in certain regions of muscle fibers.

When astrocytic processes proliferate to form scars at the surface of the brain, the processes which are separated from the pial tissue acquire an unusually large concentration of orthogonal assemblies (1). Whether this occurs in differently produced glial scars, and especially those that form in the depths of CNS, is now under investigation. The limited capacity of the mammalian CNS for regeneration after lesions may be related to this special feature and may reflect the tendency of such glial hypertrophies to rearrange various brain compartments.

A barrier to paracellular diffusion has been demonstrated to exist between endothelial cells lining blood vessels in most regions of the nervous system. The epithelial lining of the choroid plexus and the arachnoidal and perineurial mesothelia also establish effective tissue compartments, but the ependymal lining does not. Exogenous large particles, such as toxic proteins and virions, appear to have only restricted routes of paracellular diffusion into the extracellular compartment of the brain. The routes may involve entrance from peripheral organs, from the peripheral ganglia that lack an effective endothelial barrier in their capillaries (10), or through those regions of the brain in which the blood-brain is absent or defective.

Access of particles from the blood vessels in the CNS normally requires an initial intracellular route through the vascular endothelium or the epithelial lining of the choroid plexus.

Experimental studies with electron dense tracers indicate that
once these large molecules are introduced into the CSF, they
can easily diffuse into the intercellular spaces of the parenchyma in both the brain and the spinal cord. Within the white
matter, preferential diffusion takes place along fiber systems
oriented in parallel.

Myelin formation brings about the establishment of two specialized
compartments, one involving the adaxonal space (bordered by the
paranodal axo-glial junction and the inner mesaxonal tight junction of the myelin sheath) and the other involving the intramyelinic extracellular space; i.e., the thin spiral gap between
adjacent turns of the myelin lamella (bordered by the zonulae
occludentes of the myelin sheath). The nodal axolemma is excluded from these compartments. Extracellular particles cannot
gain direct access to the adaxonal space and the intramyelinic
extracellular space without a certain degree of disruption of
the myelin sheath.

Acknowledgement. Supported by National Institutes of Health
Grant 09904.

REFERENCES

(1) Anders, J.J., and Brightman, M.W. 1979. Assemblies of
     particles in the cell membranes of developing, mature and
     reactive astrocytes. J. Neurocytol. 8: 797-800.

(2) Droz, B. 1979. How axonal transport contributes to main-
     tenance of the myelin sheath. TINS 2: 146-148.

(3) Ellisman, M.H. 1979. Molecular specializations of the
     axon membrane at nodes of Ranvier are not dependent upon
     myelination. J. Neurocytol. 8: 719-735.

(4) Ellisman, M.H.; Friedman, P.L.; and Hamilton, W.J. 1980.
     The localization of sodium and calcium to Schwann cell
     paranodal loops at nodes of Ranvier and of calcium to com-
     pact myelin. J. Neurocytol. 9: 185-205.

(5) Fiori, M.G., and Mugnaini, E. 1981. Microglia-like cells
     in the chicken ciliary ganglion. Neurosci., in press.

(6)   Friedrich, V.L., Jr.; Massa, P.; and Mugnaini, E. 1980.
       Fine structure of oligodendrocytes and central myelin
       sheaths. In Search for the Causes of Multiple Sclerosis
       and Other Chronic Diseases of the CNS, ed. A. Boese,
       pp. 27-49. Weinheim: Verlag Chemie.

(7)   Henkart, J.; Landis, D.M.D.; and Reese, T.S. 1976. Simi-
       larity of junctions between plasmamembranes and endoplas-
       mic reticularic muscle and neurons. J. Cell Biol.
       70: 338-347

(8)   Henrikson, C.K., and Vaughn, J.E. 1974. Fine structural
       relationships between neurites and radial glial processes
       in developing mouse spinal cord. J. Neurocytol. 3: 659-679.

(9)   Heuser, J.E., and Salpeter, S.R. 1979. Organization of
       acetylcholine receptors in quick-frozen, deep-etched, and
       rotary-duplicated "torpedo" postsynaptic membrane. J. Cell
       Biol. 82: 150-173.

(10)   Jacobs, J.M. 1977. Penetration of systemically injected
       horseradish peroxidase into ganglia and nerves of the
       autonomic nervous system. J. Neurocytol. 6: 607-618.

(11)   Langley, O.K. 1979. Histochemistry of polyanious in
       peripheral nerve. In Complex Carbohydrates of Nervous
       Tissue, eds. R.U. Margolis and R.K. Margolis, pp. 193-208.
       New York: Plenum Press.

(12)   Madson, J.K., and Møllgård, K. 1979. The tight epithe-
       lium of the Mongolian gerbil subcommissural organ as re-
       vealed by freeze-fracturing. J. Neurocytol. 8: 481-491.

(13)   Mugnaini, E. 1978. Fine structure of myelin sheaths. In
       Proceedings of the European Society for Neurochemistry,
       ed. V. Newhoff, pp. 3-31. Weinheim: Verlag Chemie.

(14)   Mugnaini, E., and Forstrønen, P.F. 1967. Ultrastructural
       studies on the cerebellar histogenesis. I. Differentiation
       of granule cells and development of glomeruli in the chick
       embryo. Zeit. Zellforsch 77: 115-143.

(15)   Peters, A.; Palay, S.L.; and Webster, H. de F. 1976.
       The Fine Structure of the Nervous System: The Neurons
       and Supporting Cells, pp. 1-406. Philadelphia: Saunders.

(16)   Pfenninger, K.H., and Maylié-Pfenninger, M.F. 1979. Sur-
       face glycoconjugates in the differentiating neuron. In
       Complex Carbohydrates of Nervous Tissue, eds. R.U. Margolis
       and R.K. Margolis, pp. 185-192. New York: Plenum Press.

(17)   Rosenbluth, J. 1976. Intramembranous particle distribution
       at the node of Ranvier and adjacent axolemma in myelinated
       axons of the frog brain. J. Neurocytol. 5: 731-745.

(18)   Rosenbluth, J.   1979.   Aberrant axo-Schwann cell
       junctions in dystrophic mouse nerves.   J. Neurocytol.
       5: 655-672.

(19)   Schnapp, B., and Mugnaini, E.   1978.   Membrane architec-
       ture of myelinated fibers as seen by freeze-fracture.
       In Physiology and Pathobiology of Axons, ed. S.G. Waxman,
       pp. 83-123.   New York: Raven Press.

(20)   Sotelo, C., and Privat, A.   1978.   Synaptic remodeling
       of the cerebellar circuitry in mutant mice and experi-
       mental cerebellar malformations.   Acta Neuropathol.
       (Berl.) 43: 19-34.

(21)   Wiley, C.A., and Ellisman, M.H.   1980.   Rows of dimeric-
       particles within the axolemma and juxtaposed particles
       within glia, incorporated into a new model for the para-
       nodal glial-axonal junction at the node of Ranvier.
       J. Cell Biol. 54: 261-280.

(22)   Wood, J.G.; Jean, D.H.; Whitaker, J.N.; McLaughlin, B.J.;
       and Albers, R.W.   1977.   Immunocytochemical localization
       of the sodium, potassium activated ATPase in knifefish
       brain.   J. Neurocytol. 6: 571-581.

(23)   Wood, J.G., and McLaughlin, B.J.   1979.   Histochemistry
       and cytochemistry of glycoproteins and glycosaminoglycans.
       In Complex Carbohydrates of Nervous Tissue, eds.
       R.U. Margolis and R.K. Margolis, pp. 139-164.   New York:
       Plenum Press.

Neuronal-glial Cell Interrelationships, ed. T.A. Sears, pp. 57-76.
Dahlem Konferenzen 1982. Berlin, Heidelberg, New York: Springer-Verlag.

Developmental Disorders of Myelination in Mouse Mutants

A. J. Aguayo and G. M. Bray
Neurosciences Unit, Dept. of Neurology, McGill University
and The Montreal General Hospital, Montreal, Quebec H3G 1A4, Canada

Abstract. Disorders of myelination have now been recognized
in the peripheral and/or central nervous systems of several
mutant mice. The abnormalities in these mutants represent
alterations of specific stages in the normal series of interactions between axons and sheath cells. Their investigation
contributes to the study of normal axon-sheath cell relationships as well as the mechanisms responsible for each specific
disorder.

INTRODUCTION

The study of structural and functional alterations in peripheral nerves of animals with hereditary disorders of myelination not only contributes to an understanding of the mechanisms involved in these specific diseases but also provides
insights into the interdependencies that exist among neurons,
sheath cells, and the neural environment. In the present
report, we review: (a) current knowledge of the processes involved in the development of these interdependencies in normal
myelinated fibers, (b) some of the contemporary techniques used
to investigate the role of the cellular and non-cellular components of peripheral nerves, and (c) the changes in nerves of
certain animal mutants with disorders of myelination that
indicate a breakdown in axon-Schwann cell interactions. It is
anticipated that the study of these abnormalities in peripheral

nerves will contribute to the understanding of myelin disorders
in the more complex central nervous system.

DEVELOPMENT

During the development of normal peripheral nerves in rodents,
the relationship between axons and Schwann cells evolves through
several stages: (a) undifferentiated Schwann cells, which originate in the neural crest or certain ectodermal placodes, migrate
with immature axons; (b) proliferation and further migration of
Schwann cells result in the ensheathment of individual axons;
(c) the differentiation of Schwann cells leads first to the
formation of a basal lamina and eventually to myelination or
the ensheathment of Remak fibers; and finally (d) during maturation, there are increases in myelin thickness and internodal
lengths, both in proportion to changes in axon diameter (28).

The regulation of these developmental events is still poorly
understood. During early migration, undifferentiated Schwann
cells are deployed both in longitudinal rows and in radially-
oriented arrays that extend along and across bundles of axons.
The number of daughter cells and the sizes of the zones which
they occupy appear to be highly variable and are probably influenced both by the relationships of the sheath cells with
axons and also by interactions among the sheath cells themselves. During this phase of development, the stimulation
and curtailment of Schwann cell division and the death of redundant cells lead to a matching of axon and Schwann cell
populations. In vitro studies suggest that there are several
Schwann cell mitogens; neurites from autonomic and sensory
neurons (24,43) as well as homogenates of such neurites (37)
cause Schwann cell division. Although the axonal component
that stimulates Schwann cell division is sensitive to trypsin
(37), little is known about the molecular structure of this
mitogen. Extracts of pituitary gland or brain (31) or cyclic
AMP (32) also stimulate Schwann cell multiplication in vitro,
but their precise role in the nerves of living animals is unknown. The cessation of Schwann cell proliferation may also be

influenced by their differentiation, their relationship to
axons, and by contact between neighboring Schwann cells.
Finally, the death of redundant axons (15) and Schwann cells
(6) may also determine the numbers of axons and Schwann cells
in developing nerves.

The differentiation of Schwann cells to form the sheaths of
myelinated or unmyelinated fibers is controlled by axons; in
this respect undifferentiated Schwann cells are bipotential
(3,41). The mechanism by which axons influence Schwann cells
to differentiate is not clear. The suggestion that axon size
is a critical factor in the initiation of myelination has been
disputed. Although it has been postulated that the axon diameter for myelination to commence in the PNS is 1 μm, much
smaller axons become myelinated in the CNS (28), and Fraher
(14), in a detailed quantitative study of developing rat ventral roots, has shown that myelinogenesis begins in axons
ranging in diameter from 0.7 to 2.5 μm. It has also been
postulated that the neuronal regulation of myelinogenesis involves a surface membrane interaction between axons and myelinating cells (16). The demonstration that the myelin-
associated glycoprotein (MAG) is present in Schwann cells and
the periaxonal portion of PNS myelin (30) lends support to
this hypothesis. However, even if MAG or some other cell
surface component(s) mediate the interaction between axons and
myelin-forming cells, it does not exclude the possibility
that the calibre of the axons influences the initiation of
myelinogenesis; indeed, it is possible that there is a relationship between axonal diameter and the composition of their
surface membranes.

Although sheath cells in both the central and peripheral nervous systems may respond to common signals for myelinogenesis
(4,42), in vitro studies suggest that signals from axons may
be more important for the synthesis of certain myelin components by Schwann cells than by oligodendrocytes. In tissue
cultures where Schwann cells are isolated from neurites,

they stop producing immunocytochemically detectable amounts
of the major myelin proteins. In contrast, oligodendrocytes
continue to synthesize some of these proteins in the absence
of neurons (22).

During maturation of peripheral nerves, there is a fairly
precise correlation between axon diameter, myelin thickness,
and internodal length (28). In other words, the axons of
myelinated fibers seem to determine the geometry of the differentiated Schwann cell by exerting an effect upon the radial
and longitudinal growth of its plasma membrane. Conversely,
there is also evidence that myelination influences axon diameter (1,33). Thus, during development, axons and Schwann cells
increase in size, each one influencing the other. The mechanisms involved in these interactions are unknown.

The role of the microenvironment in the regulation of axon-
Schwann cell relationships has elicited considerable recent
interest (see Bunge, this volume). Fibroblasts and the extracellular matrix may be essential for the formation of Schwann
cell basal lamina and may also be involved in the process of
attachment and ensheathment of axons by Schwann cells.

METHODS USED FOR THE INVESTIGATION OF CELL INTERACTIONS IN
MOUSE MUTANTS

Nerve Transplants

When a segment of one nerve is grafted into another, the
Schwann cells from the donor nerve survive, multiply, and
ensheath the axons that regenerate from the recipient nerve
(2). Because the graft segment of such regenerated nerves
combines host axons and transplanted Schwann cells, it is
possible to study in vivo the relationships between selected
populations of axons and Schwann cells, each originating from
different experimental animals or even from experimental animal and human nerves. Such experimental nerve grafts have
been applied to the study of Schwann cell differentiation (3),
multiplication (2), and the pathogenesis of some hereditary
disorders of myelination in man (5,13,23). Schwann cells

grown in vitro may also be transplanted into peripheral nerves
(2) or into demyelinated spinal cords of mice (11). In addition,
regeneration and myelination have been studied in experimental
combinations of peripheral nerve axons and transplanted optic
nerves, (4,42), as well as after transplanting peripheral nerve
segments into transected rat spinal cords (36).

Mouse Chimeras
Experimental chimeras are produced by aggregating 8-cell, pre-
implantation, embryos (29). The resulting composite organism
contains two or more genotypically distinct cell lines that
co-exist throughout embryogenesis and in the mature animal,
thus providing a unique opportunity for the analysis of cell-
cell interactions and clonal deployment in primary development.

Tissue Culture
Techniques that permit the separation of neurites, Schwann
cells, and fibroblasts have been used to study normal myelination and pathogenetic mechanisms in nerves of myelin
deficient mutants (see Bunge, this volume).

MUTANT MICE WITH DISORDERS OF MYELINATION IN PERIPHERAL NERVES
Trembler Mice
In this dominantly-inherited disorder which results from a
mutation on chromosome 11, myelin is either abnormally thin,
poorly compacted, or totally absent in peripheral nerves
(1). In addition, there is evidence for myelin breakdown.
Because of persistent multiplication, populations of Schwann
cells in Trembler nerves are nearly ten times those of normal
controls. In contrast to these abnormalities of myelinated
fibers, the morphology and density of Schwann cells are normal
in unmyelinated (Remak) fibers (2).

Investigations of cell combinations in experimental nerve
grafts carried out in mature animals (1), in vitro cultures of
Schwann cells (9), and Trembler/normal chimeras (35) have shown
that a primary Schwann disorder is responsible for the abnormalities in Trembler nerves. In nerve graft experiments, the myelin

deficit found in Trembler nerves can be reproduced when Trembler
Schwann cells are transplanted into nerves of normal mice; conversely, in the reciprocal experiment, axons from Trembler animals can be myelinated normally by grafts of normal Schwann cells.
The heightened proliferation of Schwann cells was also reproduced
in the regenerated Trembler grafts (1). Because collagen and
other elements in the neural environment may influence myelination (8), their contribution to the pathogenesis of the Trembler
abnormality could not be excluded from the nerve graft experiments
in which collagen, fibroblasts, and perineural cells were transplanted together with Schwann cells. Tissue culture experiments
subsequently demonstrated that pure preparations of Trembler
Schwann cells when combined with normal axons, reproduce the
Trembler abnormality (9). The existence of a primary Schwann
cell deficit was further supported by the demonstration in
Trembler/normal mouse chimeras that the Trembler phenotype is
expressed even by Schwann cells which have undergone their entire
primary development in association with normal Schwann cells
and axons (35).

The Schwann cells in Remak fibers of Trembler mice are phenotypi-
cally normal but the characteristic hypomyelination and increased
Schwann cell proliferation are replicated when a nerve composed
of Remak fibers, seen as the cervical sympathetic trunk, is
transplanted from Trembler into a richly-myelinated nerve, such
as the sural nerve, in normal mice (25). Thus, although Trembler
Schwann cells carry the abnormal gene, the Trembler abnormality
is only expressed when these cells are required to form myelin.
The results of these experiments also indicate that Trembler
Schwann cells are capable of ensheathing axons in unmyelinated
fibers, but they are unable to differentiate beyond this stage
to form myelin.

Although the abnormal persistence of Schwann cell multiplication
in Trembler nerves leads to a greatly increased population of
Schwann cells, their number becomes stable approximately a month
after birth (1). The lack of further increase in the number of

Trembler Schwann cells, in spite of their continued multiplication, suggests that there is a concomitant death of Schwann
cells. One trigger for Schwann cell proliferation in these
Trembler nerves may be the breakdown products of myelin (16).
In addition, recurrent demyelination and the death of ensheathing cells may expose other Schwann cells to the putative axonal
mitogen. Finally, the inability of Trembler Schwann cells to
differentiate may make them particularly sensitive to these and
other mitogenic influences.

Transplantation experiments using Trembler and normal peripheral
nerves also demonstrate the influences of Schwann cells on
axon diameter (1). Axon calibre in unoperated Trembler nerves
is above half normal, but it returns to normal when the axons
are myelinated by grafted normal Schwann cells. Conversely,
when Trembler Schwann cells ensheath normal axons, the diameters
of the axon within the grafted segment are decreased. Thus,
axon diameter may depend on local interactions with the ensheathing cells, as well as on influences that arise within the peri-
karyon and the peripheral fields of innervation.

The Trembler neuropathy is an important example of a genetic
disorder in which the cellular interdependencies in peripheral
nerve fibers are disrupted by a primary defect of one component, the Schwann cells. Trembler Schwann cells seem to be
unable to differentiate beyond the stage of primary ensheathment
of axons to form and sustain myelin normally; Schwann cell
multiplication, Schwann cell death, and a reduction of axon
calibre are secondary manifestations (Fig. 1).

Quaking Mice
Animals affected with this disorder, which is due to an autosomal recessive mutation on chromosome 17, have a widespread
deficit of myelin in both the central and peripheral nervous
systems (38,40). The myelin sheaths in peripheral nerves of
quaking mice are abnormally thin although Schwann cells numbers are increased as a result of continuing proliferation (1).
The deficiency of myelin is much less severe than in Trembler

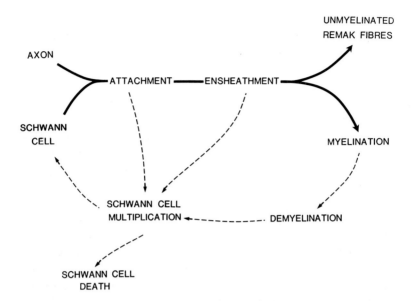

FIG.1 - Suggested sequence of changes in peripheral nerves of Trembler mice. While some Schwann cells contact and ensheath Remak fibers normally, others fail to differentiate into producing normal myelin. This failure leads to demyelination, Schwann cell multiplication, and probably to the death of many of these cells.

and only becomes apparent when a recessively thick myelin sheath is required.

In experiments using nerve grafting techniques, it has been established that there is a primary Schwann cell defect in the quaking mutant (1). However, the hypomyelination in quaking peripheral nerves is probably due to a defect that occurs later in the formation of myelin than that found in Trembler mice. Central nervous system axons become normally myelinated when Schwann cells, cultured in vitro from normal rats, are transplanted into the spinal cords of quaking mice (11). The formation of normal myelin by these xenogenic Schwann cells suggests that the CNS axons (or more general conditions) are not responsible for the myelin deficiency in the quaking central nervous system. The presence of hypomyelination in the central and peripheral nervous systems also suggests that the affected gene normally regulates myelin formation by both Schwann cells and oligodendrocytes.

Dystrophic Mice

This mutation, on chromosome 10, affects several tissues including skeletal muscle and certain peripheral nerves. In the peripheral nervous system there are both focal and generalized alterations in axon ensheathment. The most striking abnormality is the presence in the ventral and dorsal spinal roots of segments where the axons are totally devoid of Schwann cells and therefore of myelin (7). Both the longitudinal and cross-sectional distribution of the abnormally-ensheathed axons is characteristic: many axons are naked at the mid-root level but are myelinated near the spinal cord and at their exit from the vertebrae canal. In transverse sections from the maximally affected regions the naked axons are grouped in bundles with their plasma membranes closely apposed. At the periphery of these bundles of naked axons there are often small cells with scanty cytoplasm and patchy or totally absent basal lamina. These cells either marginally contact, or only partially encircle, individual axons but fail to ensheath them normally. Tritiated thymidine labelling studies of dystrophic mice have shown that these cells continue to divide throughout the life of the animal, well beyond the neonatal period of normal Schwann cell proliferation. These abnormal cells, which resemble those observed early in the development of normal peripheral nerves, have been shown to be arrested Schwann cells (26). At the same levels that one sees bundles of naked axons there are also a few myelinated fibers that are ensheathed by typical Schwann cells. The number of these fibers is strikingly smaller than normal, and they tend to be situated in the periphery of the spinal root and in the neighborhood of small blood vessels. Many myelinated fibers in spinal roots and peripheral nerves of dystrophic mice have inappropriately thin myelin sheaths, excessively wide nodes of

Ranvier, and shorter than normal internodes. All Schwann cells
in the spinal roots and peripheral nerves of dystrophic mice are
surrounded by an abnormal basal lamina which shows frequent dis-
continuities. This basal lamina defect is not present in other
tissues (20).

Although the structure of neurons and axons in dystrophic mice
tends to be similar to that of controls, the mean axon diameter
of the naked axon segments is smaller than that of the myelinated
portions of the same spinal root. These changes more likely
reflect the dependencies of axon calibre on myelination than an
intrinsic abnormality of the dystrophic roots. Although the
plasma membranes of adjacent naked axons are directly apposed and
there is ephaptic transmission of electrical impulses (34), no
specialized interaxonal contacts have been identified.

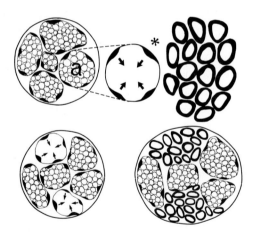

FIG. 2 - Top row (left to right). Schematic representation of
a developing nerve. Axon bundles (a) are surrounded by undifferentiated Schwann cell (*) which proliferate and migrate radially (arrows) into such bundles to ensheath and myelinate individual axons. Bottom row: In dystrophic mouse spinal roots most
of these cells fail to differentiate and migrate locally. In
contrast, some Schwann cells successfully ensheath axons in the
same segments of these roots.

Abnormal groups of naked axons are first observed in the spinal roots of dystrophic mice shortly before birth. Although the total numbers of Schwann cells in spinal roots are similar in dystrophic and control animals beyond the age of two weeks, nearly half of the Schwann cells in the dystrophic animals are undifferentiated; the failure of myelination in these nerves seems to reflect this deficit. Most of the undifferentiated cells remain in the periphery of axon bundles and fail to migrate into these bundles to ensheath individaul axons (Fig. 2).

When dystrophic spinal roots are transplanted into the peripheral nerves of normal or dystrophic mice, the degree of myelination after regeneration is indistinguishable from that produced by transplantation of normal spinal roots (1,26). In addition, the undifferentiated cells, which continue to multiply but fail to ensheath axons normally in the intact dystrophic spinal roots, can myelinate axons in these nerve grafts. Thus, the naked axons in the dystrophic spinal roots do not appear to be the result of an intrinsic defect of the local Schwann cells. Furthermore, because the continuously renewing population of undifferentiated Schwann cells in segments of the spinal roots of dystrophic mice is persistently unable to ensheath the axons of myelinated fibers, there must be an ongoing process that prevents the differentiation of the Schwann cells, not only during development but throughout the life of the animal.

In spite of persistent division of the undifferentiated cells, the total population of Schwann cells remains stable in the spinal roots of dystrophic mice beyond the age of three weeks (27). This finding suggests that, as in Trembler mice, the proliferation of undifferentiated Schwann cells is balanced by an equivalent loss. If Schwann cells in vivo have a finite potential for cell division (as is suggested by in vitro doubling experiments with other mammalian tissues (17)), it is possible that the Schwann cells in these mutants die prematurely because they have exhausted their pre-determined capacity for mitotic activity. An alternative, but less likely explanation

for the levelling out of Schwann cell numbers in dystrophic
spinal roots, is that some of the cells migrate out of the
nerve.

In contrast to the restricted location of the naked axons in
the dystrophic spinal roots, the basal lamina defects as well
as the abnormal length of some nodal gaps are more generalized
abnormalities which also affect the peripheral nerves of this
mutant. These generalized defects are reproduced in tissue
culture of dystrophic dorsal root ganglia (24) and the defective basal lamina is the only abnormality thus far demonstrated
to be reproduced in regenerated grafts of dystrophic roots transplanted into normal nerves. Although these findings suggest
that the basal lamina defect is due to a primary Schwann cell
disorder, other cellular and non-cellular influences may be involved. An axonal role is unlikely because in both the nerve
graft and tissue culture experiments, the basal lamina defect
was reproduced when Schwann cells became associated with normal
axons. However, in both types of experiment the fibroblasts and
endoneurial collagen as well as the Schwann cells were of dystrophic origin. More recently, in vitro experiments have suggested
that the basal lamina defect in vivo may be due to an aberrant
interaction between the genetically abnormal Schwann cells and
a defective component of the extracellular matrix (see Bunge).
The relationship, if any, of the basal lamina defect and the
failure of axon ensheathment by Schwann cells in the spinal roots
of dystrophic mice remains to be determined.

Shiverer Mice
In this recessively inherited disorder, myelin in the central
nervous system is poorly compacted, lacks major dense lines, and
contains no detectable levels of myelin basic protein (12). Although myelin basic protein is also absent in the peripheral
nervous system, the morphology of the peripheral myelin is essentially normal (18). Thus, although the presence of myelin
basic protein may be a prerequisite for the compaction and
stability of CNS, it probably fulfills a different function

in the peripheral nervous system. Experimental nerve grafts
of shiverer and normal mouse nerves indicate that the defect
in the synthesis of myelin basic protein is reproduced by Schwann
cells transplanted from this mutant (Sidman and Broshart, personal communication).

Twitcher Mice

Homozygous twitcher mice appear normal at birth but then lose
weight, develop a tremor, become progressively weak, and finally
die by three months of age. The central nervous system shows
demyelination, gliosis, and axonal degeneration. In peripheral
nerves, there is also demyelination and remyelination. Macro-
phage-like cells containing a variety of inclusions are seen in
both the central and peripheral nervous system (10). These
histopathologic findings as well as the demonstration that ga-
lactosyl ceramidase is deficient in these animals (19) indicate
that twitcher is the murine counterpart of the human and canine
globoid leukodystrophies.

Three months after transplantation of nerve segments from young
twitcher mice into normal litter mates, there is a replication
of the abnormalities that are characteristic of peripheral nerves
in this mutant (39). Further graft studies using the nerves from
this mutant offer promising possibilities for the investigation
of the effects of a normal enzyme source on cells bearing the
twitcher genotype.

CONCLUSION

The findings in these mutants illustrate the multiplicity of
primary and secondary alterations that may occur in peripheral
nerves; these include failures of Schwann cell deployment, proliferation, differentiation and survival, defective formation
of basal lamina, demyelination, and changes in the size and
properties of axons. The study of these alterations not only
provides clues for the understanding of the mechanisms governing the establishment and maintenance of axon-Schwann cell relationships in normal myelinated fibers but also indicates that

the timing, intensity, and localization of these changes influence the specific functional and histological features characteristic of these and other peripheral neuropathies (Fig. 3).

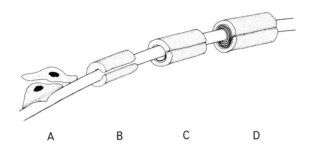

A   B   C   D

FIG. 3 - Schematic illustration of the development of axon-
Schwann cell relationships in a myelinated peripheral nerve
fiber. Multiplying Schwann cells initially contact axons (A),
which they encircle by the elaboration of cellular processes
(B). Myelination (C) follows primary ensheathment and continues until the thickness of the myelin sheath is appropriate
to the calibre of the axon (D). In spinal roots of dystrophic
mice, Schwann cells do not evolve beyond the stage (A). In
Trembler nerves most fibers remain at (B) while in quaking
there is a failure to evolve from (C) into (D).

In dystrophic mice, axonal segments remain totally unensheathed
because their early association with Schwann cells is prevented.
In Trembler mice, primary ensheathment is possible but myelin
formation is defective. In quaking, the initial stages of axon
ensheathment and myelin formation are normal but maturation of
the myelin sheath is impaired. In shiverer, there is a qualita-
tive difference in peripheral myelin as the result of the defective production of a specific protein, and in twitcher a
recently-identified lysosomal enzyme defect causes acute myelin
breakdown and may damage axons.

In addition to providing a dynamic view of the responses of
peripheral nerve components, the findings reviewed here have
also documented that specific cellular or non-cellular components

of peripheral nerve can be primary targets of the genetic abnormality. In Trembler and quaking mice, a primary, intrinsic
Schwann cell defect interferes with normal myelination, while in
dystrophic mice, factors extrinsic to the Schwann cell itself
may continuously prevent axonal ensheathment by Schwann cells.

REFERENCES

(1)   Aguayo, A.J.; Bray, G.M.; and Perkins, C.S.   1979.   Axon-
      Schwann cell relationships in neuropathies of mutant mice.
      Ann. N.Y. Acad. Sci. 317: 512-531.

(2)   Aguayo, A.J.; Bray, G.M.; Perkins, C.S.; and Duncan, I.D.
      1979.   Axon-sheath cell interactions in peripheral and
      central nervous system transplants. Soc. Neurosci. Symp.
      4: 361-383.

(3)   Aguayo, A.J.; Charron, L.; and Bray, G.M. 1976. Potential
      of Schwann cells from unmyelinated nerves to produce myelin:
      a quantitative ultrastructural and autoradiographic study.
      J. Neurocytol. 5: 565-573.

(4)   Aguayo, A.J.; Dickson, R.; Trecarten, J.; Attiwell, M.; Bray,
      G.M.; and Richardson, R. 1978.   Ensheathment and myelina-
      tion of regenerating PNS fibers by transplanted nerve glia.
      Neurosci. Lett. 9: 97-104.

(5)   Aguayo, A.J.; Perkins, C.S.; Duncan, I.; and Bray, G.M. 1978.
      Human and animal neuropathies studied in experimental nerve
      transplants. In Peripheral Neuropathies, eds. N. Canal and
      N. Pozza, pp.37-48. Amsterdam: Elsevier/North Holland Bio-
      medical Press.

(6)   Berthold, C.H. 1973.   Local demyelination in developing fe-
      line nerve fibers. Neurobiol. 3: 339-352.

(7)   Bradley, W.G., and Jenkison, M. 1973. Abnormalities of
      peripheral nerves in murine muscular dystrophy. J. Neurol.
      Sci. 18: 227-247.

(8)   Bunge, R.P., and Bunge, M.B. 1978. Evidence that contact
      with connective tissue matrix is required for normal inter-
      action between Schwann cells and nerve fibers. J. Cell Biol.
      78: 943-950.

(9)   Cornbrooks, C.; Cochran, M.; Mithen, F.; Bunge, M.B.; and
      Bunge, R.P. 1980. Myelination in explants of Trembler dor-
      sal root ganglion grown in culture. Trans. Am. Soc. Neuro-
      chem. 11: 110.

(10) Duchen, L.W.; Eicher, E.M.; Jacobs, J.M.; Scaravilli, F.; and
      Teixeira, F. 1980. A globoid cell type of leucodystrophy in
      the mouse: The mutant Twitcher. In Neurological Mutations
      Affecting Myelination. Research Tools in Neurobiology, ed.
      N. Baumann, pp. 107-114. Amsterdam: Elsevier/North Holland.

(11) Duncan, I.D.; Aguayo, A.J.; Bunge, R.P.; and Wood, P.M. 1980.
      Transplantation of rat Schwann cells grown in tissue culture
      into the mouse spinal cord. J. Neurol. Sci., in press.

(12) Dupouey, P.; Jacque, C.; Bourre, J.M.; Cesselin, F.; Privat, A.; and Baumann, N. 1979. Immunochemical studies of myelin basic protein in shiverer mouse devoid of major dense line of myelin. Neurosci. Lett. $\underline{12}$: 113-118.

(13) Dyck, P.J.; Lais, A.C.; Sparks, M.F.; Oviatt, K.F.; Hexum, L.A.; and Steinmuller, D. 1979. Nerve xenografts to apportion the role of axon and Schwann cell in myelinated fiber absence in hereditary sensory neuropathy, type II. Neurology (Minneap.) $\underline{29}$: 1215-1221.

(14) Fraher, J.P. 1972. A quantitative study of anterior root fibers during early myelination. J. Anat. $\underline{112}$: 99-124.

(15) Fraher, J.P. 1974. A numerical study of cervical and thoracic ventral nerve roots. J. Anat. $\underline{118}$: 127-142.

(16) Hall, S.M. 1978. The Schwann cell: A reappraisal of its role in the peripheral nervous system. Neuropath. and Appl. Neurobiol. $\underline{4}$: 165-176.

(17) Hayflick, L. 1975. Current theories of human aging. Fed. Proc. $\underline{34}$: 9-13.

(18) Kirschner, D.A., and Ganser, A.L. 1980. Compact myelin exists in the absence of basic protein in the shiverer mutant mouse. Nature $\underline{283}$: 207-210.

(19) Kobayashi, T.; Scaravilli, F.; and Suzuki, K. 1980. Biochemistry of twitcher mouse: an authentic murine model of human globoid cell leukodystrophy. In Neurological Mutations Affecting Myelination. Research Tools in Neurobiology, ed. N. Baumann, pp. 253-256. Amsterdam: Elsevier/North Holland.

(20) Madrid, R.E.; Jaros, E.; Cullen, M.J.; and Bradley, W.G. 1975. Genetically determined defect of Schwann cell basement membrane in dystrophic mouse. Nature $\underline{257}$: 319-321.

(21) McCarthy, K.D., and Partlow, L.M. 1976. Neuronal stimulation of (^{3}H)-thymidine incorporation by primary cultures of highly purified non-neuronal cells. Brain Res. $\underline{114}$: 415-426.

(22) Mirsky, R.; Winter, J.; Abney, E.R.; Pruss, R.M.; Gavrilovic, J.; and Raff, M.C. 1980. Myelin specific proteins and glycolipids in rat Schwann cells and oligodendrocytes in culture. J. Cell Biol. $\underline{84}$: 483-494.

(23) Ohnishi, A.; Tateishi, J.; Matsumoto, T.; et al. 1979. Fabry disease: cellular expression of enzyme deficiency in nerve xenografts. Neurology (Minneap.) $\underline{29}$: 899-901.

74   A.J. Aguayo and G.M. Bray

(24) Okada, E.; Bunge, R.P.; and Bunge, M.B. 1980. Abnormalities
     expressed in long-term cultures of dorsal root ganglia from
     the dystrophic mouse. Brain Res. 194: 455-470.

(25) Perkins, C.S.; Aguayo, A.J.; and Bray, G.M. 1981. Behaviour
     of Schwann cells from Trembler mouse unmyelinated fibers
     transplanted into myelinated nerves. Exp. Neurol., in press.

(26) Perkins, C.S.; Bray, G.M.; and Aguayo, A.J. 1980. Evidence
     for undifferentiated Schwann cells in the spinal roots of
     dystrophic mice. Can. J. Neurol. Sci. 7: 123.

(27) Perkins, C.S.; Bray, G.M.; and Aguayo, A.J. 1980. Persis-
     tent multiplication of axon associated cells in spinal roots
     of dystrophic mice. Neuropath. Appl. Neurobiol. 6: 83-91.

(28) Peters, A.; Palay, S.L.; and Webster, H.deF. 1976. The
     Fine Structure of the Nervous System, 2nd ed., p. 406.
     Philadelphia: Saunders.

(29) Peterson, A.C.; Frair, P.M.; Rayburn, H.R.; and Cross, D.P.
     1979. Development and disease in the neuromuscular system
     of muscular dystrophic<-->normal mouse chimeras. Soc.
     Neurosci. Symp. 4: 258-273.

(30) Quarles, R.H.; McIntyre, L.J.; and Sternberger, N.H. 1979.
     Glycoproteins and cell surface interactions during myelino-
     genesis. Soc. Neurosci. Symp. 4: 322-343.

(31) Raff, M.C.; Abney, E.; Brockes, J.P.; and Hornby-Smith, A.
     1978. Schwann cell growth factors. Cell 15: 813-822.

(32) Raff, M.C.; Hornby-Smith, A.; and Brockes, J.P. 1978. Cy-
     clic AMP as a mitogenic signal for cultured rat Scwhann cells.
     Nature 273: 672-673.

(33) Raine, C.S.; Wisniewski, H.; and Prineas, J. 1969. An ultra-
     structural study of experimental demyelination and remyelina-
     tion. II. Chronic experimental allergic encephalomyelitis in
     the peripheral nervous system. Lab. Invest. 21: 316-327.

(34) Rasminsky, M. 1978. Ectopic generation of impulses and
     cross-talk in spinal nerve roots of "dystrophic" mice. Ann.
     Neurol. 3: 351-357.

(35) Rayburn, H.; Peterson, A.C.; and Aguayo, A.J. 1980. Develop-
     ment and deployment of Schwann cells in peripheral nerves of
     Trembler <--> normal mouse chimeras. Soc. Neurosci. Abst.,
     VI: 660.

(36) Richardson, P.M.; McGuinness, U.M.; and Aguayo, A.J. 1980.
     Axons from CNS neurons regenerate into PNS grafts. Nature
     284: 264-265.

(37) Salzer, J.L.; Glaser, L.; and Bunge, R.P. 1977. Stimulation of Schwann cell proliferation by a neurite membrane fraction. J. Cell Biol. 75: 118a.

(38) Samorajski, T.; Friede, R.L.; and Reimer, P.R. 1970. Hypomyelination in the quaking mouse. A model for the analysis of disturbed myelin formation. Neuropath. Exp. Neurol. 29: 507-523.

(39) Scaravilli, F.; Jacobs, J.M.; and Teixeira, F. 1980. Quantitative and experimental studies on the twitcher mouse. In Neurological Mutations Affecting Myelination, Research Tools in Neurobiology, ed. N. Baumann, pp. 115-122. Amsterdam: Elsevier/North Holland.

(40) Sidman, R.L.; Dickie, M.M.; and Appel, S.H. 1964. Mutant mice (quaking and jumpy) with deficient myelination in the central nervous system. Science 144: 309-311.

(41) Weinberg, H.J., and Spencer, P.S. 1976. Studies on the control of myelinogenesis. II. Evidence for neuronal regulation of myelin production. Brain Res. 113: 363-378.

(42) Weinberg, H.J., and Spencer, P.S. 1979. Studies on the control of myelinogenesis. III. Signalling of oligodendrocyte myelination by regenerating peripheral axons. Brain Res. 162: 273-279.

(43) Wood, P.M., and Bunge, R.P. 1975. Evidence that sensory axons are mitogenic for Schwann cells. Nature 256: 662-664.

Neuronal-glial Cell Interrelationships, ed. T.A. Sears, pp. 77-91.
Dahlem Konferenzen 1982. Berlin, Heidelberg, New York: Springer-Verlag.

The Development of Electrical Excitability

N. C. Spitzer
Biology Dept., University of California, San Diego
La Jolla, CA 92093, USA

Abstract. During the embryonic development of several different types of nerve cells there are significant changes in the
ionic dependence of the inward current of the action potential.
In addition, voltage-dependent uncoupling has recently been
observed between developing neurons. During axonal regeneration, excitability develops in both old and new membrane;
development of excitability in the newly-formed membrane may
recapitulate earlier embryonic development. The role of excitability in development remains enigmatic; the available
experimental evidence suggests that the Na^+ action potential
plays little or no role in early neural development.

INTRODUCTION

It has recently become possible to obtain rather detailed

information about the development of electrical excitability

in individual nerve cells. This is a substantial advance be-

yond our previous knowledge about the onset of excitability

in whole organisms, or even in particular tissues. It seems

likely that this advance will permit the analysis of the con-

tributions of excitability to neural development, of the role

of inexcitable cells, and of the molecular events that under-

lie neuronal maturation and the interaction between neurons

and their surrounding cellular and ionic environments.

This paper focuses chiefly on the embryological development

of electrical excitability in neurons. To date, most atten-

tion has been directed toward the development of the action

potential mechanism. More recently another form of electri-

cal excitability has been discovered, the presence of voltage-

dependent uncoupling between developing neurons. The salient
features of this phenomenon and its possible relationship to
the embryonic action potential are discussed. There is also
substantial interest in the development of excitability in
regenerating nerve cells, both with respect to alterations in
the excitability of the preexisting neuronal membranes and
the development of excitability in newly formed membrane. The
possible roles of electrical excitability in neuronal development are also of interest, as is the contribution of neighboring glial cells. Some aspects of this work have been recently reviewed (14, 35).

There have been three general approaches to the study of the
development of electrical excitability. The first has involved the impalement of cells with one or more microelectrodes, to clamp the voltage of the cell membrane at different
levels and examine the ensuing ionic currents, or to depolarize the cell briefly (current clamp) and record the change in
membrane potential resulting from the ionic currents. The
latter procedure is technically easier but usually less in-
formative since the voltage change is not a direct measure of
ionic current when the membrane resistance is changing. By
altering the ionic environment and using specific pharmacological blocking agents, it has been possible to identify the
ions that are involved and to examine the kinetic properties
of the ion channels. Collectively, these approaches have been
the most useful to date. Unfortunately they are difficult to
apply to most embryonic cells which are small, fragile, and
often hard to identify. Furthermore, the low resting potentials of damaged cells can cause the inactivation of ion
channels that may then be interpreted to be absent. The second approach has been to use drugs that hold open particular
channels for prolonged periods (e.g., veratridine for the
voltage-dependent Na channel); the resulting continuous ion
fluxes can then be detected fairly easily, by the accumulation of radioisotopes and other procedures. This technique
is valuable when the cells are uniformly sensitive to the

drug throughout development. The third approach involves the
specific binding of probes which have a high affinity for
specific ion channels, such as radiolabeled saxitoxin for the
Na^+ channel. This method yields information about the number
and distribution of channels, without microelectrode impalements, but one cannot draw inferences about the ion fluxes
through the channels, and changes in binding may occur independently of changes in permeability.

THE DEVELOPMENT OF THE ACTION POTENTIAL MECHANISM

The present evidence indicates that there is often a difference between the ionic dependence of the action potential as
it first appears during development and that expressed in
mature cells. Furthermore, this change frequently involves the
gradual conversion of Ca^{++} dependent impulses into those dependent on Na^+. These conclusions are drawn from studies in
which it has been possible to make intracellular recordings
from embryonic cells at very early stages and raise the
question of the function of different types of excitability
during development.

Neurons in the Embryo

The investigation of the Rohon-Beard neurons in the spinal
cord of frog embryos illustrates the general features of these
studies (2,37). They are primary sensory neurons that have
their birthdate during the gastrula stage (22). These cells
are electrically inexcitable when examined just prior to clo-
sure of the neural tube. Within a few hours, however, they
are able to produce impulses that consist of an overshooting
plateau of long duration, often several hundred msec. The
overshoot increases with increases in Ca^{++} concentration in
the saline but is unaffected by removal of Na^+ or addition of
tetrodotoxin (TTX). In contrast, agents that block Ca^{++}
channels, such as Co^{++}, La^{+++}, or Mn^{++}, block these events.
Thus, the inward current is carried chiefly by Ca^{++}. At a
little over a day of embryonic development, the appearance of
the action potential changes and begins to consist of a spike

followed by a plateau; it is now only tens of msec in duration. The same tests indicate that the spike is due to an
inward Na$^+$ current, while the plateau is Ca^{++}-dependent.
Finally, by 3 to 4 days of development the action potential
has the appearance of impulses in many mature excitable cells and
consists of a slender spike about one msec in duration (see
Figure l). Now the overshoot increases with increases in the
Na$^+$ concentration and Ca^{++} channel blockers have little effect.
The action potential has come to depend largely on an influx

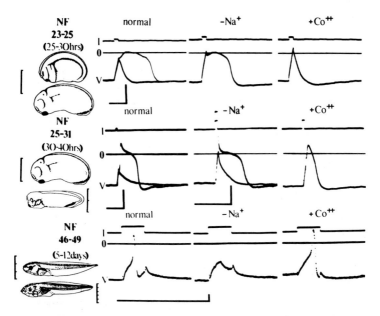

FIG. 1 - Action potentials from three different Rohon-Beard
neurons typical of different periods of development in vivo.
The size of the embryos is indicated in mm. The calibration
scale is 20 msec. 40 mV, and 5 nA (2 nA at bottom). Top row:
at early stages the action potential, when elicited, is of
long duration. It is largely unaffected by replacment of Na$^+$
with Tris but abolished by the addition of Co^{++} to normal
saline. Middle row: the action potential elicited by a sufficiently large current pulse consists of a spike followed by
a plateau. The spike is eliminated by removal of Na$^+$ (a larger current is needed to elicit the plateau). The plateau is
in turn eliminated by Co^{++}. Bottom row: the action potential
is blocked by removal of Na$^+$ but largely unaffected by the
presence of Co^{++}.

of Na^+. Thus the ionic dependence of the inward current of
these action potentials changes during development, from Ca^{++}
to Na^+. The dorsal root ganglion cells develop after the
Rohon-Beard neurons in the same frog embryos and are thought
to take over the sensory function of this latter population
when it disappears. These cells exhibit a spectrum of electrical excitability that suggests a similar sequence of development (1). Transection of the olfactory nerve in adult bull-
frogs produces degeneration of the neurons and the generation
of a new population. The action potentials in the new, outgrowing axons are Ca^{++}-dependent, while those in the mature
axons depend on Na^+ (40).

Neurons in the grasshopper embryo have also proven accessible
to microelectrode impalements at very early stages, permitting
analysis of development of the action potential mechanism by
the techniques described above (10,11). Ca^{++}/Na^+-dependent
action potentials are elicited from the cell bodies when these
cells first become excitable, and either ion alone is sufficient to support an impulse. Later both ions are required to
produce an action potential in the adult. Thus there appears
to be a change in the ionic mechanism of the action potential
during development. In contrast there is no developmental
change in the Na^+-dependent action potentials arising in the
axons.

Neurons in Culture

In a number of instances it has been possible to make cultures
of dissociated cells from regions of the embryonic nervous
system and study the properties of the cells in vitro. This is
a particularly attractive experimental approach, since development in vitro often parallels development in vivo, and it
renders the neurons even more accessible for experimental manipulation. Cultures of the neural plate and underlying meso-
derm of frog embryos have permitted further study of the development of the action potential mechanism (38,42). Analysis of
birthdate and the kind of chemosensitivity shown by these cells

suggests that some of them are the sensory Rohon-Beard neurons,
while others are likely to be motor neurons. The processes
of the neurons are clearly visible in those low-density cultures, although they can only rarely be discerned in dissected preparations in vivo, and the cultures are prepared at a
sufficiently early stage that these are primary rather than
regenerating neurites. These neurons have Ca^{++}-dependent
action potentials both in their cell bodies and processes,
that are converted to Na^+-dependent action potentials over a
period of several days, paralleling development in vivo. Since
they develop normally while not contiguous with other neurons
or nonneuronal cells, cell contacts are unnecessary during
this period of differentiation.

The differentiation of the action potential mechanism of mouse
neuroblastoma cells also follows the sequence from Ca^{++} to Na^+
under some conditions (26). Studies of neurons in vitro have
often been conducted on cultures after they have been established for several weeks. The ionic dependence of impulses
in mouse dorsal root ganglion cells appears to shift from Ca^{++}/
Na^+ to TTX-sensitive Na^+-dependent potentials (24), but in
other cases the ionic basis of action potentials did not
change during the period of study. Neurotransmitters or their
blocking agents can modulate voltage-dependent ion channels
(5,8,15,20). In some cases the transmitters have been shown
to act on Ca^{++} channels. Analysis of the development of this
interesting relationship may illuminate our understanding of
the development of electrical excitability.

Other Cell Types
Action potentials can be produced by the eggs of some animals,
indicating that embryonic development must involve the loss as
well as the acquisition of ion channels, since not all cells
are electrically excitable. One role of voltage-dependent ion
channels in egg membranes may be to provide a block to poly-
spermy (12). The development of striated muscle cells has
been examined in the chick; leg muscle cells are initially

inexcitable one week before hatching, but develop a Ca^{++}
action potential that is gradually converted to a Na^+ action
potential by one week after birth (17). Tunicate striated
muscle cells have Ca^{++}/Na^+ action potentials at early stages
of development that later come to depend on Ca^{++} alone (27,
41). In the development of a clonal line of rat skeletal
muscle cells, myoblasts produce small Na^+ action potentials;
after fusion, the myotubes produce Ca^{++}/Na^+ action potentials
(18,19). Cardiac muscle cells of the chick heart ventricle
have Ca^{++}/Na^+ action potentials at early stages of embryonic
development that later depend on Na^+ alone. This shift in
ionic dependence has also been seen in cells grown in culture,
under some conditions. A similar change in the ionic dependence of the action potential occurs in the development of the
rat heart (23).

THE DEVELOPMENT OF VOLTAGE-DEPENDENT UNCOUPLING

A different form of electrical excitability has been recently
described in developing embryos. Rohon-Beard neurons, like
many other embryonic cells and like other developing neurons
(10,11,31,34), are electrically coupled at very early stages,
before they can make action potentials (36). This can be
demonstrated by injecting current into one cell and observing
that this changes the membrane potential not only of that
cell but also of other nearby cells. The current presumably
spreads via gap junctions, and since it passes equally well
in both directions, the junctions are of the non-rectifying
variety. Polarization of the cells by current pulses that
are of sufficient amplitude and duration may cause the cells
to become relatively uncoupled. After a latent period the
voltage change in the injected cell increases, and there is
a concomitant decrease in the voltage change in the second
cell (Figure 2A, B). This phenomenon is called voltage-
dependent uncoupling, since the degree of coupling depends
on the voltage difference between the cells. The coupling
coefficient drops to 10% of its initial value when the voltage difference between the two cell bodies is 70-90 mV. Cells

often become rapidly recoupled (within seconds) after the
current pulse, and the phenomenon can be repeated for as long
as the impalements are maintained. Voltage-dependent uncoupling does not involve chemical synapses since it persists in
the presence of Co^{++} ions and is unaffected by depolarization
of the cells with high K^+. This is not a global phenomenon
since other unidentified cells in the spinal cords of the same
embryos show electrical coupling that is not voltage-dependent.

Within a few hours Ca^{++} action potentials can be elicited from
these cells, while they are still coupled. An impulse in one
occasionally causes a sufficient depolarization of an adjacent
one to initiate a second impulse there. More frequently, a
Ca^{++} action potential in one cell causes depolarization of a
second cell that is subthreshold for impulse initiation. However the Ca^{++} action potential in the first cell can be of
sufficient size and duration to cause the coupling coefficient
to decrease during its course. The Rohon-Beard cells become
completely uncoupled at about the time the Na^+ component of
the action potential appears and they remain uncoupled during
further development. Other cells in the spinal cord continue
to show voltage-dependent uncoupling; these may be other
neurons at an early stage of their development.

The voltage-dependence of electrical coupling was first described for isolated pairs of amphibian blastomeres (39). Voltage clamp analysis has shown that the junctional conductance
decreases as a function of transjunctional voltage. The movement of Lucifer Yellow Dye between coupled cells is greatly
reduced when the cells are relatively uncoupled by transjunctional polarization. These findings suggest that voltage-
dependent uncoupling may play an important role in the regulation of intercellular communication during early neural
development.

THE ACTION POTENTIAL MECHANISM DURING REGENERATION
Damage to the nervous system, when not overwhelming, is often

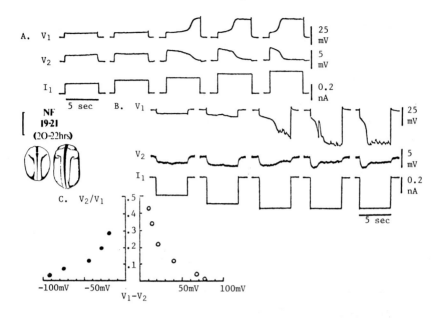

FIG. 2 - Voltage-dependent uncoupling between two different
pairs of Rohon-Beard neurons at very early stages of development. The size of the embryo is indicated in mm. A. Depolarizing current pulses of increasing amplitude and constant
duration cause an abrupt increase in the voltage recorded in
the first cell (V_1), concomitant with a decrease in the voltage recorded in the second cell (V_2). The coupling coefficient
(V_2/V_1) decreases. The latency of this uncoupling decreases
as the current strength is increased. B. The same phenomenon
is observed with hyperpolarizing current pulses, for a different pair of cells. The cells recouple rapidly at the end
of the current pulses. C. Plot of the coupling coefficient
as a function of the difference between the membrane potentials of the two cell bodies, at the end of the current pulses
(o; A; ●; B). The coupling coefficient falls to about 10%
with (V_1-V_2) of 70-90 mV.

followed by a process of regeneration. The effects of acute
injury on the excitability of intact portions of neurons have
been studied in some vertebrate and invertebrate preparations.
Following transection of ventral roots, cat motoneurons show
dendritic action potentials (6, 21) which appear to be due to
the appearance of new ionic channels in pre-existing dendritic

membrane. Cutting the axons of certain neurons in the insect
nervous system or exposing them to colchicine causes an
altered excitability of the cell body that seems to be due to
the appearance of Na^+ channels in pre-existing membrane (9,
29). However, regeneration also involves the extension of
new processes and the elaboration of new membrane, so two
questions arise: when does this new membrane become excitable
and what kind of excitability does it have? Cultures of
dissociated dorsal root ganglion cells from adult guinea pigs
have been examined; the neurites were truncated by the dissection and dissociation process and then regrew in vitro (7).
Na^+ action potentials and Ca^{++} action potentials could be
elicited by varying the ionic environment. The mean rate of
rise of the Na^+ action potentials did not change with the age
of the cultures. But the mean rate of rise of the Ca^{++} action
potentials increased significantly during the period of initial, active neurite outgrowth. Although several interpretations of this result are possible, the findings are consistent
with the presence of Ca^{++} action potentials in the membrane
of the new neurites that contribute to the Ca^{++} action potential recorded in the cell body. Recordings from regenerating
giant axons of adult cockroaches, close to the site of transection (25), revealed that the membrane was initially inexcitable and later produced Ca^{++} action potentials that were
gradually converted to the normal Na^+ action potentials. These
findings suggest that the development of excitability during
regeneration parallels the sequence seen in embryonic development.

THE ROLE OF EXCITABILITY IN DEVELOPMENT
It is far from clear what role, if any, excitability may play
in development, beyond the simple expression of an important
neuronal phenotype. In principle, it is possible that Ca^{++}
action potentials of long duration may serve to deliver pulses
of Ca^{++} to the interior of the cells - which could be important for such events as exocytosis, the activation of intracellular contractile mechanisms, or the alteration of

intracellular pH or cyclic nucleotide levels. Another possibility is that by reversing the sign of the membrane potential for a relatively prolonged period, Ca^{++} action potentials promote the insertion of other membrane components. The observation that Ca^{++} action potentials cause uncoupling between developing amphibian neurons also suggests the possibility that these potentials may restrict intercellular communication between cells that have different developmental fates.

A developmental role for excitability can be excluded when it appears relatively late. The embryonic grasshopper neurons first show action potentials when their axons arrive at their peripheral targets (10). Development and regeneration of axons appears to be unaffected by levels of TTX that block Na^+ action potentials (13,14,28,32); the possibility that Ca^{++} action potentials are present and playing some role is more difficult to eliminate. There is some evidence that long-term blockage of action potentials with TTX, thereby blocking synaptic connections and possibly trophic interactions, is toxic to cells with Na^+ spikes (3). Genetic tools are being developed that should permit more detailed analysis of this issue. In Drosophila the "Shaker" and the temperature-sensitive paralytic "No-action potential" mutants probably have defective voltage-dependent K^+ and Na^+ channels, respectively (16,43). The existence of these mutants means that it should be possible to determine the effects of the absence of different kinds of electrical excitability during development. Such studies have already been carried out with another temperature-sensitive paralytic mutant, "Shibire" (4,30,33), but the nature of the underlying defect in this mutant is less clear. At present the available evidence suggests that Na^+ action potentials are not necessary for neuronal development.

Acknowledgements. The author is supported by grants from the ONR and the NIH.

88

N.C. Spitzer

REFERENCES

(1) Baccaglini, P.I. 1978. Action potentials of embryonic dorsal root ganglion neurones in Xenopus tadpoles. J. Physiol. 283: 585-604.

(2) Baccaglini, P.I., and Spitzer, N.C. 1977. Developmental changes in the inward current of the action potential of Rohon-Beard neurones. J. Physiol. 271: 93-117.

(3) Bergey, G.K.; Fitsgerald, S.C.; Schrier, B.K.; and Nelson, P.G. 1980. Neuronal maturation in mammalian cell culture is dependent on spontaneous electrical activity. Brain Res., in press.

(4) Buzin, C.H.; Dewhurst, S.A.; and Seecof, R.L. 1978. Temperature-sensitivity of muscle and neuron differentiation in embryonic cell cultures from the Drosophila mutant, shibire^{ts1}. Dev. Biol. 66: 442-456.

(5) Dunlap, K., and Fischbach, G.D. 1978. Neurotransmitters decrease the calcium component of sensory neurone action potentials. Nature 276: 837-839.

(6) Eccles, J.C.; Libet, B.; and Young, R.R. 1958. The behaviour of chromatolysed motoneurones studied by intracellular recording. J. Physiol. 143: 11-40.

(7) Fukuda, J., and Kameyama, M. 1979. Enhancement of Ca spikes in nerve cells of adult mammals during neurite growth in tissue culture. Nature 279: 546-548.

(8) Gartner, T.K.; Land, B.; and Podleski, T.R. 1976. Genetic and physiological evidence concerning the development of chemically-sensitive voltage-dependent ionophores in L6 cells. J. Neurobiol. 7: 537-549.

(9) Goodman, C.S., and Heitler, W.J. 1979. Electrical properties of insect neurones with spiking and nonspiking somata: normal, axotomized and colchicine-treated neurones. J. Exp. Biol. 83: 95-121.

(10) Goodman, C.S., and Spitzer, N.C. 1979. Embryonic development of identified neurones: differentiation from neuroblast to neurone. Nature 280: 208-214.

(11) Goodman, C.S., and Spitzer, N.C. 1981. The development of electrical properties of identified neurones in grasshopper embryos. J. Physiol., in press.

(12) Hagiwara, S., and Jaffe, L.A. 1979. Electrical properties of egg cell membranes. Ann. Rev. Biophys. Bioeng. 8: 385-416.

(13) Harris, W.A. 1980. The effect of eliminating impulse activity on the development of the retino-tectal projection in salamanders. J. Comp. Neurol. 194: 303-317.

(14) Harris, W.A. 1981. Neural activity and development. Ann. Rev. Physiol. 43: 689-710.

(15) Horn, J.P., and McAfee, D.A. 1980. Alpha-adrenergic inhibition of calcium-dependent potentials in rat sympathetic neurones. J. Physiol. 301: 191-204.

(16) Jan, Y.N.; Jan, L.Y.; and Dennis, M.J. 1977. Two mutations of synaptic transmission in Drosophila. Proc. Roy. Soc. Lond. B. 198: 87-108.

(17) Kano, M. 1975. Development of excitability in embryonic chick skeletal muscle cells. J. Cell Physiol. 86: 503-510.

(18) Kidokoro, Y. 1973. Development of action potentials in a clonal rat skeletal muscle cell line. Nature New Biol. 241: 158-159.

(19) Kidokoro, Y. 1975. Sodium and calcium components of the action potential in a developing skeletal muscle cell line. J. Physiol. 244: 145-159.

(20) Klein, M., and Kandel, E.R. 1978. Presynaptic modulation of voltage-dependent Ca current: mechanism for behavioural sensitization in Aplysia californica. Proc. Nat. Acad. Sci. 75: 3512-3516.

(21) Kuno, M., and Llinas, R. 1970. Enhancement of synaptic transmission by dendritic potentials in chromatolysed motoneurones of the cat. J. Physiol. 210: 807-821.

(22) Lamborghini, J.E. 1980. Rohon-Beard cells and other large neurons in Xenopus embryos originate during gastrulation. J. Comp. Neurol. 189: 323-333.

(23) Lieberman, M., and Sano, T. 1976. Developmental and Physiological Correlates of Cardiac Muscle. New York: Raven Press.

(24) Matsuda, Y.; Yoshida, S.; and Yonezawa, T. 1978. Tetrodotoxin sensitivity and Ca component of action potentials of mouse dorsal root ganglion cells cultured in vitro. Brain Res. 154: 69-82.

(25) Meiri, M.; Spira, M.E.; and Parnas, I. 1981. Membrane conductance and action potential of a regenerating axonal tip. Science, in press.

(26) Miyake, M. 1978. The development of action potential mechanism in a mouse neuronal cell line in vitro. Brain Res. 143: 349-354.

(27) Miyazaki, S.; Takahashi, K.; and Tsuda, K. 1972. Calcium and sodium contributions to regenerative responses in the embryonic excitable cell membrane. Science 176: 1441-1443.

(28) Obata, K. 1977. Development of neuromuscular transmission in culture with a variety of neurons and in the presence of cholinergic substances and TTX. Brain Res. 119: 141-153.

(29) Pitman, R.M. 1975. The ionic dependence of action potentials induced by colchicine in an insect motoneurone cell body. J. Physiol. 247: 511-520.

(30) Poodry, C.A.; Hall, L.; and Suzuki, D.T. 1973. Developmental properties of shibire[ts]: a pleiotropic mutation affecting larval and adult locomotion and development. Dev. Biol. 32: 373-386.

(31) Rayport, S.C., and Kandel, E.R. 1980. Developmental modulation of an identified electrical synapse: functional uncoupling. J. Neurophysiol. 44: 555-567.

(32) Roper, S., and Ko, C.-P. 1978. Impulse blockade in frog cardiac ganglion does not resemble partial denervation in changing synaptic organization. Science 202: 66-68.

(33) Salkoff, L., and Kelly, L. 1980. Aminopyridines mimic mutant Drosophila developmental defects. Comp. Biochem. Physiol. 65: 59-63.

(34) Sheridan, J.D. 1978. Junction and experimental modification. In Intercellular Junctions and Synapses, eds. J. Feldman, N.B. Gilula, and J.D. Pitts. London: Chapman and Hall.

(35) Spitzer, N.C. 1979. Ion channels in development. Ann. Rev. Neurosci. 2: 363-397.

(36) Spitzer, N.C. 1980. Electrical uncoupling of vertebrate spinal cord neurons during development. Abstract. Society of Neurosciences, 10th Annual Meeting, p. 287.

(37) Spitzer, N.C., and Baccaglini, P.I. 1976. Development of the action potential in embryonic amphibian neurons in vivo. Brain Res. 107: 610-617.

(38) Spitzer, N.C., and Lamborghini, J.E. 1976. The development of the action potential mechanism of amphibian neurons isolated in culture. Proc. Nat. Acad. Sci. 73: 1641-1645.

(39) Spray, D.C.; Harris, A.L.; and Bennett, M.V.L. 1979. Voltage-dependence of junctional conductance in early amphibian embryos. Science 204: 432-434.

(40) Strichartz, G.; Small, R.; Nicholson, C.; Pfenninger, K.H.; and Llinas, R. 1980. Ionic mechanisms for impulse propagation in growing nonmyelinated axons: saxitoxin binding and electrophysiology. Abstract. Society of Neurosciences, 10th Annual Meeting, p. 660.

(41) Takahashi, K.; Miyazaki, S.; and Kidokoro, Y. 1971. Development of excitability in embryonic muscle cell membranes in certain tunicates. Science $\underline{171}$: 415-418.

(42) Willard, A.L. 1980. Electrical excitability of outgrowing neurites of embryonic neurones in cultures of dissociated neural plate of Xenopus laevis. J. Physiol. $\underline{301}$: 115-128.

(43) Wu, C.-F., and Ganetzky, B. 1980. Genetic alteration of nerve membrane excitability in temperature-sensitive paralytic mutants of Drosophila melanogaster. Nature $\underline{286}$: 814-816.

Group on Ontogeny

Standing, left to right: Albert Aguayo, Guy McKhann, Louis Lim,
Nick Spitzer, Hans Distel, Enrico Mugnaini.
Seated: Michael Rickmann, Gary Strichartz, Henry de F. Webster,
Max Cowan, Pasko Rakic.

Neuronal-glial Cell Interrelationships, ed. T.A. Sears, pp. 93-114.
Dahlem Konferenzen 1982. Berlin, Heidelberg, New York: Springer-Verlag.

Ontogeny
State of the Art Report

G. R. Strichartz, Rapporteur
A. J. Aguayo, W. M. Cowan, H. Distel, L. Lim,
G. M. McKhann, E. Mugnaini, P. Rakic,
M. J. Rickmann, N. C. Spitzer, H. de F. Webster

INTRODUCTION

There are a number of processes which are important for the
attainment of normal interactions between neurons and glial
cells[*] during development. Achieving the appropriate numbers
of different types of glial cells and their correct proportion
to neurons, the deployment of glia and neurons to their proper
destinations in the central (CNS) and peripheral nervous systems
(PNS), the induction of glial process outgrowth, axonal ensheathment, and the initiation and termination of myelin formation,
are all critical processes for the formation of the properly
functioning brain. To simplify this total process we have subdivided the cellular events into several stages and considered
each stage in regard to the specific involvement of different
cell types and to the interactions among different cells.

---

[*] The terms glia and glial are used in this report to cover all
varieties of central glial cells (astrocytes, oligodendrocytes,
and microglia) and peripheral supporting cells such as Schwann
cells and the satellite cells found in sensory and autonomic
ganglia.

Although we have emphasized the formation of myelin during development, it seems clear that many of the processes occurring during this time are also probably relevant to the mechanism of remyelination. Finally, insofar as the Conference as a whole focused on neuronal-glial cell relationships with regard to multiple sclerosis, and recognizing that this disease is essentially one of myelin breakdown and is not a primary neuronal defect, this report particularly emphasizes the role of those glial cells involved in forming myelin. Relatively little attention is paid to the developmental aspects of neurons; these are reviewed by Spitzer (this volume) and in a recent article by Cowan (3).

VARIETIES OF GLIAL CELLS

Any discussion of the development of the nervous system must begin with a classification of the various types of cells involved. We have focused exclusively on the varieties of glia, on the reasonable assumption that variations in myelination and remyelination are more a reflection of generalized rules governing neuron-glia interactions throughout the nervous system than an expression of the properties of particular neuronal cells in specific regions of the nervous system.

Traditionally, several types of glial cells have been recognized in the vertebrate nervous system. During development, and in response to local injury, certain glial cells either appear transiently (radial glia), increase their number (microglia), or become hypertrophic (astrocytes). Both protoplasmic and fibrillary astrocytes are found throughout the central nervous system. Typically these display many slender radiating processes. In the CNS interfascicular oligodendrocytes are the cells which form myelin sheaths. Perineuronal or satellite oligodendrocytes surround cell bodies in the CNS; it is not known whether they have the potential to form myelin. In the peripheral nervous system myelin is formed by Schwann cells. Axons of small diameter (<1.0 m) are ensheathed by Schwann cells but do not become myelinated. Satellite glial cells are found in sensory and

autonomic ganglia of the PNS and a separately differentiable
set of myenteric glia is found in the plexuses of the gut.

The glial cells identified earliest in development by their
morphology and GFA (glial fibrillary acidic protein) immuno-
staining are the radial glia first recognized by Cajal and
described by Rakic elsewhere in this volume. Glial cell classes
in the adult CNS that possess radial processes which are similar
in their morphology and antigenic properties are the Müller
cells (retina) and the Bergmann glia (cerebellum). Derivatives
of primitive radial glia include fibrous and protoplasmic
astrocytes. Features identifying oligodendrocytes as a separate
class of CNS glia vary during development. Their origin and
the time at which they appear in relation to neuronal and astroglial differentiation remain poorly understood. In part this
is because at early stages they appear to share some morphological features (size, nuclear shape, chromatin density and
distribution, endoplasmic reticulum, polysomes, microtubules)
with some neuronal precursors, and in part because the presently
available immunochemical markers for oligodendrocytes only appear relatively late in development. In contrast to Schwann
cells in the PNS, oligodendrocytes generally only ensheath
axons that become myelinated; in addition they have not been
shown to secrete collagen nor to form a basal lamina.

Astrocytes often separate developing neurites into large bundles,
and oligodendrocytes are found early in association with astro-
citic processes as axon ensheathment and CNS myelination begin.
As this occurs, numerous junctional complexes appear between
adjacent astrocytes and oligodendrocytes (see Mugnaini, this
volume). These interactions may be important in oligodendrocyte-
axon contact, ensheathment, and the onset of myelination. As
myelination begins, the myelin associated glycoprotein (MAG)
and myelin basic protein (BP) can be detected in oligodendroglia (25,26).

The origin of microglia is uncertain, but available evidence
suggests that they are derived from blood-borne phagocytes which

invade the developing nervous system. When nerve terminals,
axons, and cell bodies degenerate, microglia become activated,
divide, and help to remove the debris. The role of these cells
in remodeling developing myelin and removing myelin breakdown
products from demyelinating lesions is unclear. Since both astrocytes and hematogenous cells participate in myelin breakdown
and phagocytosis, a specific microglia role in this process
cannot be defined.

The Schwann cells, which are of placodal, neural crest, and neural
tube origin, are glial cells in the PNS which form myelin and
also ensheath neurons and unmyelinated axons, No morphological,
functional, or antigenic properties have been identified that
distinguish placodally derived Schwann cells from those that
originate from the neural crest. Although Schwann cells and
oligodendrocytes both form myelin, they do so by different strategies, described in more detail below.

Criteria for Glial Classification

There are several criteria for identifying and discriminating
among different glial types. The first group of criteria is
morphological and includes the shape of the cells, the appearance of their plasma membrane specializations, and a variety of
cytoplasmic inclusions and organelles. The second set of criteria rest on the staining properties of the cells; different
types of glia are stained selectively by certain histological
procedures, e.g., Cajal's gold sublimate method for astrocytes
and Hortegas' silver carbonate method for microglia. The third
set of criteria are based on specific cellular antigens, located on surface membranes or in the cytoplasm (see Schachner,
et al., this volume). Currently there are several antigenic
markers specific for astroglia and oligodendrocytes, and one
(RAN 1) which selectively recognizes Schwann cells. Astrocytes
contain the well characterized GFA protein, the function of which
is unknown. Vimentin is another protein found in astrocytes and
ependymal cells (but not oligodendrocytes); it polymerizes to
form intermediate filaments, whose roles have not yet been defined. While several selective stains and some specific antigenic

markers are available, there are really very few markers which
are completely specific for any one cell strain, and this deficiency is especially serious when one considers the development
of glia. Thus it is impossible at present to get at precisely
when specific cell types appear and to suggest, even on temporal
evidence, the stem cells from which they originated.

Glia can also be classified according to their functions, although
these are often only inferred by appearance and prove difficult
to assess directly. It appears that glia may be specialized to
perform many functions, all in support of primary neural activity.
They provide structural support for the guidance and/or translocation of cells during development and act as a mechanical support
matrix throughout life. They may also provide some of the metabolic or biosynthetic requirements of neurons; this has been demonstrated for ensheathing cells of large invertebrate axons where
an exchange of locally synthesized proteins between glia cells
and the axons occurs (12). A transfer of phospholipids from
Schwann cells to axons has also been detected in the PNS (7).
The phagocytic action of astroglia is important for the removal
of cellular debris in response to injury or disease and, transiently, during neural death in development. Enzymatic activites
of glia are probably important both in normal function and in
response to injury or disease. Glia actively remove neurotransmitter compounds from intercellular spaces in the brain and may
actively take up ions such as K^+, which accumulate intercellularly
as a result of nerve activity, although the latter function has
been called in question recently (see Orkand, this volume). To
our knowledge, the electrical activity of glial cells is quite
different from that of most excitable cells. Their membranes
are selectively permeable to potassium ions, and changes in glial
membrane potential in response to alterations of extracellular
potassium ions are markedly different than those of nerve membranes. Furthermore, agents which activate sodium channels in
neurons, thus depolarizing these cells, do not affect glia, and
radiolabelled markers for sodium channels do not specifically
bind to glial membranes (30).

The major problem with using the positive criteria listed above
for identifying glial cells is that most, if not all, of them
may change during development. Changes in the morphology function, and even in immunological markers have been reported, as
the cells migrate and mature in embryonic and postnatal tissue,
and as they respond to injury or disease. At present the antigenic markers are the most sensitive and the most specific for
recognizing cell types and the production and use of new antibodies should help to elucidate the ontogenesis and distribution
of the various glial cell types during development.

Another potentially useful, and as yet almost unexploited, approach is the analysis of nucleic acids, including messenger
RNA, which must differ among cell types as early as the first
differentiation from stem cell precursors.

GENERAL STRATEGY OF MYELIN FORMATION
The proper development of the nervous system requires a systematic,
sequential organization of cellular differentiation, migration,
proliferation, specialization (both within individual cells and
among groups of cells), and the termination of certain cellular
events, some of which may lead to the selective death of cells.
The remainder of this chapter briefly summarizes one aspect of
our limited knowledge of these processes.

In myelin formation, the events of importance are staged throughout neural development. Certain glial cells arise from precursor
cells within the ventricular zone of the neural tube and form a
scaffolding for the migration of other glia and neurons. While
the earliest glia disappear during later development, the later-
appearing astroglia and oligodendroglia proliferate, interacting
with each other and with developing neurons. At some stage following their disposition along the course of growing axons, oligodendrocytes begin to extend cytoplasmic processes towards nerve
fibers and, having contacted and ensheathed them, begin the process of myelin formation. Later, myelin formation ceases and
the relatively stable myelin sheath is maintained at constant

diameter for most of the life of healthy individuals, although
there may be some loss of myelin thickness in old age. In the
following sections we discuss the origins of glial cells, their
migration and proliferation, and the specialized structures and
features of glia, including their interactions with each other
and with neurons. We have concentrated on the myelination processes and particularly on the signal(s) which initiate and
regulate oligodendrocyte extension, myelin formation, and the
termination of myelination.

ORIGINS OF GLIAL CELLS
In discussing the origin of glial cells, we must consider separately the origin of central and peripheral glia. It appears
that peripheral glia (Schwann cells and ganglionic satellite cells
and myenteric glia) originate from three sources: the neural
crest is the major source, and the neural tube and certain placodes (epithelial formations derived from ectoderm, such as the
olfactory and otic placode) are minor sources. Recent studies
indicate that there are also peripheral microglia of undetermined origin which resemble ultrastructually their CNS counterparts (9). Oligodendrocytes, which will form myelin in the CNS,
are probably derived from the neuroepithelium of the ventricular
zone lining the lumen of the neural tube.

The variety of glial cell types causes difficulties when the
divergence of the astrocytic and oligodendrocytic cell lines is
investigated. Glial precursors have been classified and described mainly during the ontogenesis of rodents by means of
light-, electronmicroscopy, and autoradiography with ^{3}H-thymidine
(for review see (18,19)). The most immature group of glial cells
cannot be described exclusively as precursors of either astrocytes or oligodendrocytes. The smallest presumptive glial cell
is the least differentiated member of this group. It shows a
dark nucleus surrounded by a small amount of cytoplasm which
contains many ribosomal rosettes and few other organelles. The
more mature presumptive glial cells have a lighter overall appearance and show a distinct nucleolus and a variety of cytoplasmic organelles. Because of methodological restrictions,

however, it remains controversial which of the presumptive
glial cell types give rise to astrocytes and which to oligodendrocytes (23,28). This is true of all the macroglia except
those derived from the radial glial cells which are considered
a special class of astrocytes (Rakic, this volume). These
elongated cells span the entire width of the cerebral wall and
in primates are distinguishable from neural precursors at a relatively early stage by the presence of GFA within their cytoplasm
(13). In rodents these cells may acquire GFA at later developmental stages than in primates (Schachner et al., this volume).
In most mammals radial glial cells disappear by transformation
into protoplasmic and fibrillary astrocytes, Bergmann glial cells
of the cerebellum, Müller cells of the retina, tanycytes of the
hypothalamus, septal cells of the medulla and spinal cord and
ependymal cells (22).

It should be emphasized that many of these cells appear during
the course of development earlier than is generally recognized
and, indeed, may be produced concomitantly with the genesis of
neurons in the same structure (Rakic, this volume). They can
be detected using several glial markers (Schachner et al., this
volume). As far as the origin of oligodendrocytes is concerned,
we do not have similarly elaborate and detailed knowledge. Cell
markers for early oligodendrocytic forms are now being sought
and a few of them have already been identified in tissue cultures
of dispersed embryonic cells, but so far they have not been
tested in situ. It is possible that oligodendrocyte precursors
originate from neuroepithelial cells in the ventricular zone and
from cells in the subventricular zone that are GFA negative in
primates (Rakic, this volume). There is an urgent need for
markers to detect this class of embryonic cells, since the presently available methods are inadequate to determine when and where
they arise.

Certain phagocytic cells have a different origin. During development or after injury to the CNS, vascular leakage provides a pathway for macrophages to enter the CNS or to migrate to the site of

injury. However, in response to neural or myelin breakdown,
even where the vasculature is intact, microglia can proliferate
in situ, in response to some, as yet unidentified, mitogenic stimulus. At present, it appears that microglia may be of both neuroectodermal and hematogenous origin.

DEPLOYMENT OF GLIAL CELLS

Our knowledge concerning the mechanisms and timing of glial cell
deployment is quite incomplete. While it is known that all neurons migrate from the ventricular zone to the region where they
will ultimately reside, the migration of most glial cells has
not been mapped, largely because of the absence of specific cell
markers for early glial cells or their precursors. It appears
that only a few precursor cells of each type actually migrate,
because proliferation of most glial cells occurs at their final
destination; any marker to detect migrating cells would, for this
reason, have to be quite sensitive and specific. Despite the
absence of detailed knowledge at this time, some general state-
ments may be noted. Glial deployment and proliferation occur
at several stages directly following glial differentiation from
stem cells. After radial glial cells have established processes
radiating through the developing neural tissue, post-mitotic
neurons migrate out to occupy positions in the developing brain.
The mechanism for guidance during this migration is unknown,
although migrating cells are probably in direct contact with the
membranes of radial glia. Before reaching the terminus of the
radial glia extensions the migrating cells become detached (20).
Astrocytes in various brain structures are provided by the translocation and transformation of radial glial cells (22). It is
not known how oligodendrocytes reach their destinations within
the white matter. The potential myelin forming oligodendrocytes
appear to be present among the axons at the time that they form
synaptic connections, but whereas ensheathment may have occurred,
myelination per se probably does not begin before the onset of
function.

In the peripheral nervous system there does not appear to be
any equivalent to the radial glial cell of the CNS. Schwann

cells migrate in the company of growing axons and appear to
be grouped in orderly longitudinal and radial arrays. Apparently,
they are able to migrate locally away from the guiding axon, although most stay in contact. Schwann cells cease dividing when
they begin forming myelin. In the CNS, oligodendrocytes probably
continue to divide while forming myelin (29).

The signals which cue and target migration of prospective oligodendrocytes in the CNS are not known. Conceivably, these may
provide potentially therapeutic factors which could be manipu-
lated to direct myelin-competent oligodendrocytes to a region of
demyelination.

THE CONTROL OF GLIAL CELL PROLIFERATION

On the reasonable assumption that the number of glial cells of
each type normally present in the mature CNS is optimal, in the
sense that it is closely related to the functional or structural
requirements of the nervous system, it would appear that there
are rather finely-tuned mechanisms that regulate the proliferation (and possibly the selective death) of each variety of glial
element.

Almost all the evidence for proliferation of specific glial cells
derives from autoradiographic experiments utilizing ^{3}H-thymidine
uptake. However, such evidence must be viewed cautiously because
^{3}H-thymidine can be incorporated in cells in the absence of cell
division. Other processes, such as cellular repair, synthesis
of mitochondria, and changes in the free thymidine pool can
produce apparently increased thymidine uptake with no real increase in cell numbers.

The primary regulatory mechanism must operate at a very early
stage, presumably either at the medullary plate stage, or in the
ventricular zone of the early neural tube. Regulation at this
stage must determine the numbers of stem cells, for each stem
cell type is appropriate for the numbers of neuronal precursors
(and possibly the numbers of other glial cell precursors) generated in the region. The nature of this mechanism and how it

operates is wholly unknown. Conceivably it may involve "signalling" by way of the gap junctions that link all cells at these
early stages. Of particular importance at these stages is the
establishment of the relative number of the so-called radial
glial cells which seem to play such an important role in the
early migration and assembly of most parts of the CNS. The recent immunocytochemical work of Rakic and his colleagues suggests
that these cells are not only present at an early stage, but also
that they form a substantial proportion of the cells in the developing ventricular zone.

Since as yet there are no specific cytochemical markers for the
other classes of glial cell precursor, we do not know when they
first appear, nor do we have any ideas as to their relative proportions. It is generally believed that they arise relatively
late, when neuronogenesis is already well advanced, or essentially
complete, but the evidence for this view is indirect and confused
by the fact that gliogenesis generally continues for some time
after neuronal proliferation has ceased. The radial glial cells
are now known to undergo a second phase of proliferation, quite
late in development. This is accompanied by the withdrawal of
the cells' luminal and subpial processes, their rounding up, and
repeated division to give rise to a substantial proportion of the
resident astrocytes in most regions of the CNS. The stimulus for
this second proliferative phase, and the factors that regulate
its frequency and duration, are unknown.

One region in which glial cells (and certain small neurons) continue to be generated throughout development, and often well into
postnatal life, is the subventricular (or subependymal) zone.
The cells in this zone appear to leave the ventricular zone at
an early stage and continue to divide long after mitosis has
ceased in the ventricular zone. We do not know what types of
glial precursors are present in this region; it is generally assumed that it can give rise to both astrocytes and oligodendrocytes, but there is no direct evidence for this. If there are
"undetermined" glial elements, it seems likely that many of

them may arise in the subventricular zone. The application of
various glial-specific markers to the analysis of the subventricular zone promises to be the most useful next step in this
field, and, in the context of this meeting, one of considerable
importance given that the lesions in multiple sclerosis are so
frequently found in various periventricular regions.

We know as little about the general developmental factors that
regulate the numbers of glia of various types as we know about
the control of various neuronal populations. The general assumption is that whatever affects the proliferation of any
given neuronal precursor population will affect the related
glial in a proportional way. One confirmation of this occurs
in the adult PNS where a predictable and constant relationship
between neuronal soma volume and the number of adjacent satellite
cells has been established in the dorsal root ganglion (2).
However, in other systems the actual evidence for this is weak:
it derives largely from the observation that all cell proliferation in the developing nervous system is highly patterned, both
in space and time, and that the numbers of glia of each type are
(presumably) appropriate for each region.

There are two specific conditions that are of interest during
glial development: (a) the ingrowth (or removal) of axons, and
(b) naturally-occurring neuronal death. De Long and Sidman (6)
have reported that there is a marked reduction in ^{3}H-thymidine
uptake in the superior collicus of mice after eye removal in the
immediate postnatal period. Since neuronal proliferation is
over by this time, this suggests that the failure of optic nerve
fibers to grow into the collicus (and/or the degeneration of
fibers that had already reached it) causes a reduction in glial
cell proliferation. Similarly, Currie and Cowan (5) have found
that after removal of the optic cup in frogs, there is a late
reduction in cell proliferation in the optic tectum which particularly affects the division of glial cells. In neither of
these studies was the type of glial cell affected identified.
Although large numbers of neurons in most parts of the developing nervous system degenerate at about the time the population

as a whole forms its connections, this massive destruction of
neurons is apparently not accompanied by a gliosis. Presumably,
the existing phagocytic mechanisms are adequate to deal with the
neuronal debris without provoking the genesis of new glia.

GLIAL PROLIFERATION IN THE ADULT NERVOUS SYSTEM
There is little evidence for the continued proliferation of
myelinating glial cells in the normal adult nervous system, although this is a matter of some theoretical interest, and significant practical importance in demyelinating disorders such as
multiple sclerosis. Is there a residual population of cells
that, under the appropriate circumstances, can proliferate to
form new myelin-forming cells? Incorporation of ^{3}H-thymidine
in rodent brain suggests that both small and more mature presumptive glial cells, and even adult astrocytes, may be able to
proliferate (11,16,23). Some immature forms of oligodendrocytes
are labelled by thymidine in 90 day old rats (11), and even
oligodendrocytes that are forming myelin appear capable of
dividing (29).

In other highly specialized tissues such as muscle, where it is
known that mature muscle cells have lost their capacity for proliferation, there remains a population of so-called satellite
cells that can be activated in injured muscle to form myoblasts
that divide repeatedly, then fuse to form young myotubes. The
population of non-specialized glial cells identified by Vaughan
and others (31) and termed by them the "third glial cell type,"
has been regarded as such a source of proliferative cells.
Others, however, believe that the cells are a form of microglia
and that while they may proliferate they do so only preparatory
to phagocytosing neuronal breakdown products. There is no evidence, at present, that they can give rise to oligodendrocytes
or form myelin. To observe the consequences of their proliferation we need to conduct double-labelling experiments to show
that such precursor cells do proliferate and then proceed to
synthesize myelin-specific gene products.

After severe lesions of the CNS there is a rapid increase in
the number of glial cells both at the site of the lesion and
along the course of the fibers arising in, passing through,
or terminating at, the injured region. If the vascular system
has been damaged it is likely that a majority of the proliferating cells is hematogenous. For less severe lesions where the
vasculature remains intact, the origin of the proliferating
cells is less certain. In zones of axon termination (when
myelin breakdown is not a factor), the best evidence is that
the dividing cells (identified by ^{3}H-thymidine uptake) are
microglia, by the usual morphological criteria. The antece-
dents of these cells are not known, nor have they been characterized by specific cytochemical markers. In regions of neuronal
degeneration and of myelin breakdown it is generally believed that
the resident astrocytes also divide. Although remyelination by
proliferating glial cells can occur after some forms of experimental demyelination (10,14), it has not been proven that mature
oligodendrocytes are responsible; the remyelination glia may
have arisen from a pool of undifferentiated glial cells that
persist in adult rodents (17). The evidence for this is not
wholly convincing and is mainly based on the finding of increased
numbers of glial cell nuclei, without specific cytochemical
identification or analysis by thymidine uptake. There is no
convincing evidence that mature oligodendrocytes proliferate
under these circumstances; but this should be re-examined by
double-labelling procedures using thymidine uptake and oligo-
dendrocyte-specific antibodies.   There is similarly little
reliable evidence about cell proliferation in regions of demyelination, although this is an issue of fundamental importance
in multiple sclerosis.

What factors have been identified as stimulants to glial proliferation? Several Schwann cell mitogens have been detected in
in vitro studies of the developing PNS (see Bunge, this volume).
These include neurites, neurite homogenates from sensory and
autonomic neurons, cholera toxin, and at least one macromolecule
derived from brain tissue. In the presence of serum, Schwann

cell ensheathment requires the presence of fibroblasts or their
products, and it seems probable that circulating factors may
complicate the process of myelination in vivo well beyond that
observed in defined medium in vitro. Several other factors
which are mitogenic for glial cells in adult mammalian tissue
have been isolated and characterized. These are presented
briefly by Walicke et al. (this volume).

MYELINATION

Once glial cells have migrated to the correct location and
proliferated to the proper density, myelination normally commences (32). In the CNS, most, if not all, of the oligodendrocytes form myelin, each cell extending in several directions to
myelinate central axons. The exception is in certain nuclei
where a few oligodendrocytes are found capping synaptic complexes
at cell bodies without forming myelin. In the PNS myelin is
the product of Schwann cells, each of which forms only one internodal segment around only one axon. Schwann cells in the PNS
also ensheath smaller, non-myelinated fibers, and until myelination commences there is no marker for discriminating two classes
of Schwann cells, thus implying that the signal for initiating
myelination requires axonal contact.

The formation of myelin results from a complex biochemical and
anatomical process. The events involved in myelin formation
must occur in specified sequence and the failure of any one of
them may prevent myelin formation. We believe that the following
sequential steps are critical:
1) A specific contact between glial cells and the axons.
2) The maturation of the myelin-forming cell to the stage where
   it is competent to form myelin. At this stage (in the CNS)
   it already expresses myelin-specific components such as
   myelin basic protein and myelin-associated glycoprotein
   (25,26).
3) The activation of the myelin-forming cell, perhaps by a
   signal from the adjacent axon(s).

4) Glial ensheathment of the axon, growth of the spirally wrapped mesaxon and subsequent compacting of the myelin.

5) Growth of the sheath, including increases in longitudinal dimension and in diameter as the axon grows.

6) The maintenance of the myelin-axon assembly, with the strong probability that structural and inductive materials are transferred between axon and sheath.

At the onset of myelination several protein components specific for myelin first appear, indicating that this specialization selectively induces gene transcription and that the process involves more than a response of the existing membrane and cytoplasmic components. In addition to a myelin-competent glial cell, a myelin-inducing axon must also be present. Such neurites may be more important for myelin formation by Schwann cells than by oligodendrocytes; in tissue culture the former cells alone stop producing myelinic proteins whereas the latter still synthesize some of these in the absence of other cells (see Bunge, this volume). Axon diameter is related to myelin formation. In the mature mammalian PNS, few axons of diameter less than 1.0μm are myelinated, whereas in the CNS myelinated axons of 0.2-0.8 microns are frequently found (1). Myelination commences while axons are still growing and the internodes lengthen as the axon diameter increases. This apparent synchrony probably manifests an intimate, bidirectional interaction between axons and glia. In the CNS the orchestration of such events is more impressive, for there a single oligodendrocyte can be in the process of simultaneously myelinating many axons at different stages of myelination. Such a cell must be regulating its membrane synthesis at different rates in different directions to be able to remain in synchrony with many growing axons.

Nerve impulses are being conducted along axons at the time of their myelination, but may be irrelevant for myelin formation since this occurs normally in tissue culture in the presence of high concentrations of tetrodotoxin, a drug which blocks normal, sodium-dependent action potentials (4). However, in embryonic nerve cell bodies and neurites (Spitzer, this volume) and in

growing tips of vertebrate non-myelinated axons (27), there are
tetrodotoxin-insensitive, calcium-dependent action potentials.
These may persist in embryonic tissues chronically exposed to
tetrodotoxin (15) and could subserve impulse propagation in the
tissue-cultured cells noted above. The phenomenon deserves
further investigation, in which electrical excitability is monitored periodically during myelin formation in vitro.

With respect to myelin initiation in the PNS, it should be noted
that an axon is first ensheathed by a Schwann cell before myelin
formation begins. The attachment points which guide Schwann cell
ensheathment may be determined by the axon itself. Intramembrane
particles from freeze-fracture experiments appear in axons at
regions where paranodal attachments will occur, even before
myelin is formed (8). Electrophysiological experiments suggest
an intriguing parallel. In peripheral axons four days after
acute demyelination by lysolecithin, local "hot spots" of inward
current are detected during impulse propagation (Bostock and
McDonald, this volume). This evidence for the clustering of
functional sodium channels may be directly related to the aggregation of intramembranous particles, but at present we lack the
information to correlate these two membrane components. However,
if initial myelination progresses similarly to remyelination,
then sodium channels clustered together by self-aggregation
or by underlying cytoskeletal structures may provide structural
signals to trigger myelin formation. A different possibility
is that clusters of sodium channels terminate the extension of
the myelin internodal segment by preventing the longitudinal
movement of other intramembranous particles, to which the outer-
most paranodal folds are joined. Many morphological specializations of the axon never seen in non-myelinated fibers appear
at nodes of Ranvier. The reciprocal nature of the interaction
of axons and oligodendrocytes or Schwann cells is undeniable,
but the nature and mechanism of these interactions remains
obscure.

EXPERIMENTAL DISORDERS OF MYELIN FORMATION
Several mutant mouse strains manifest disorders of PNS myelin
formation, and these are described in detail by Aguayo and
Bray (this volume). The results of transplant, tissue culture,
and chimera experiments generally indicate that disorders of
myelin formation arise from defects in Schwann cell expression
involving the formation of myelin (trembler and quaking mutants)
or the establishment of basement membrane, which just precedes
myelin formation (dystropic mouse), or the proliferation and
survival of Schwann cells. Axon defects do not appear to contribute to these mutations, although secondary consequences,
such as abnormally narrow axon diameters, do appear in abnormally
myelinated fibers. Aguayo and Bray's paper documents the importance of several different factors acting at different phases
to accomplish normal myelin formation and provides a dissection
of the dynamic aspects of myelin formation as revealed by genetic
pathology.

Finally, the role of cells other than Schwann cells and oligodendrocytes and the effects of serum factors on the formation
and regulation of myelin should be addressed. Myelination by
Schwann cells in vitro is reviewed by Bunge (this volume), and
he discusses the importance of axonal mitogens for the proliferation of Schwann cells and the necessity of secretory activity
by Schwann cells for their full functional expression. The
interrelationships between Schwann cells and fibroblasts in the
presence of serum suggest that myelin-forming cells may require
the presence of other cells for their expression in vivo.
Junctional complexes are observed between astrocytes and oligodendrocytes at the time when CNS myelination begins and electrical coupling between glial cells during development have been
correlated with the presence of morphologically defined junctional complexes. The importance of these connections between glial
cells is undetermined, but since there are several experimental
methods for uncoupling electrically-coupled cells, (21,24) the
experimental investigation of the question is possible.

RELATIONSHIP OF DEVELOPMENT TO REMYELINATION

Our discussion revealed that the factors controlling myelination
in vivo were still largely undetermined. Comparison of the
processes of myelination and remyelination are likewise limited
largely by our ignorance. Nevertheless, the results of tissue
culture and transplant experiments provide some grounds for
optimism that myelination by locally delivered normal Schwann
cells in demyelinated peripheral plaques may provide a useful
therapy. The situation in the CNS is more difficult, as it
requires the delivery of normal oligodendrocytes to distributed plaques at a time when their original route and mechanism
of transport has long disappeared. Some form of directed migration of oligodendrocytes to plaques may be possible, but the
means are not apparent at this time. But, if oligodendrocytes
which had already formed myelin could be induced to begin
dividing again, they might provide a new set of glia to myelinate existing and potential plaques throughout the CNS. The
potential of mature oligodendrocytes for division and myelination
deserves to be investigated.

REFERENCES

(1) Berthold, C.-H. 1978. Morphology of normal peripheral
     axons. In Physiology and Pathobiology of Axons, ed. S.G.
     Waxman, pp. 3-63. New York: Raven Press.

(2) Biunchi, R.; Culligaris, B.; Ventura, R.; and Weibel, E.R.
     1972. Quantitative relationships between nerve and
     satellite cells in spinal ganglia. An electron micro-
     scopical study. I. Mann. Uals. Brain Res. 46: 215-234.

(3) Cowan, W.M. 1978. Aspects of neural development. Int.
     Rev. Physiol. 17: 149-191.

(4) Crain, S.M.; Bornstein, M.B.; and Peterson, E.R. 1968.
     Maturation of embryonic CNS tissues during chronic exposure
     to agents which prevent bioelectric activity. Brain Res.
     8: 363-372.

(5) Currie, J., and Cowan, W.M. 1974. Some observations on
     the early development of the optic tectum of the frog
     (Rana pipiens), with special reference to the effects of
     early eye removal on mitotic activity in the larval tectum.
     J. Comp. Neurol. 156: 123-142.

(6) Delong, G.R., and Sidman, R.L. 1962. Effects of eye re-
     moval at birth on histogenesis of the mouse superior
     colliculus: an autoradiographic analysis with tritiated
     thymidine. J. Comp. Neurol. 118: 204-225.

(7) Droz, B.; DiGamberrardino, L.; Koenig, H.L.; Boyenval, J.;
     and Hassig, R. 1978. Axon-myelin transfer of phospholipid
     components in the course of their axonal transport as
     visualized by radio-autography. Brain Res. 155: 347-353.

(8) Ellisman, M.H. 1979. Molecular specializations of the axon
     membrane at nodes of Ranvier are not dependent upon myelina-
     tion. J. Neurocytol. 8: 719-735.

(9) Fiori, M.G., and Mugnaini, E. 1981. Microglia-like cells
     in the chicken ciliary ganglion. Neuroscience, in press.

(10) Herndon, R.M.; Price, D.L.; and Weiner, L.P. 1977. Re-
     generation of oligodendroglia during recovery from de-
     myelinating disease. Science 195: 693-694.

(11) Kaplan, M.S., and Hinds, J.W. 1980. Gliogenesis of astro-
     cytes and oligodendrocytes in the neocortical grey and
     white matter of the adult rat: Electron microscopic analy-
     sis of light radioantographs. J. Comp. Neurol. 193: 711-
     727.

(12) Lasek, R.J.; Gainer, H.; and Barker, J.L. 1977. Cell to
     cell transfer of glial proteins to the squid giant axon.
     The glia to neuron protein transfer hypothesis. J. Cell.
     Biol. 74: 501-523.

(13)   Levitt, P.R., and Rakic, P.   1980.   Immunoperoxidase
       localization of glial fibrillary acid protein in the
       embryonic rhesus monkey.   J. Comp. Neur.   193: 817-848.

(14)   Ludwin, S.K.   1979.   An autoradiographic study of cellu-
       lar proliferation in remyelination of the central nervous
       system.   Am. J. Pathol.   95: 683-696.

(15)   McDonald, T.F.; Sachs, H.G.; and DeHaan, R.L.   1973.
       Tetrodotoxin desensitization in aggregates of embryonic
       chick heart cells.   J. Gen. Physiol.   62: 286-302.

(16)   Mori, S., and Leblond, C.P.   1970.   Electron microscopic
       identification of three classes of oligodendrocytes and
       a preliminary study of their proliferative activity in
       the corpus callosum of young rats.   J. Comp. Neur. 139:
       1-30.

(17)   Patterson, J.A.; Privat, A.; Ling, E.A.; and Leblond, C.P.
       1973.   Investigation of glial cells in semi-thin sections.
       III. Transformation of subependymal cells into glial
       cells, as shown by radioautography after ^{3}H-thymidine in-
       jection into the lateral ventrical of the brain of young
       rats.   J. Comp. Neurol.   149: 83-102.

(18)   Privat, A.   1975.   Postnatal gliogenesis in the Mammalian
       Brain.   In International Review of Cytology, eds. G.H.
       Bourne and J.F. Danielli, vol. 40, pp. 281-324.   New York:
       Academic Press.

(19)   Privat, A., and Fulcrand, J.   1978.   Neuroglia - from the
       subventricular precursor to the mature cell.   In Cell,
       Tissue and Organ Cultures in Neurobiology, eds. S. Federoff
       and L. Hertz, pp. 11-37.   New York: Academic Press.

(20)   Rakic, P.   1972.   Mode of cell migration to the superfi-
       cial layers of fetal monkey neocortex.   J. Comp. Neur.
       145: 61-84.

(21)   Rose, B., and Lowenstein, W.R.   1975.   Permeability of
       cell junction depends on local cytoplasmic calcium activ-
       ity.   Nature (London) 254: 250-252.

(22)   Schmechel, D.E., and Rakic, P.   1979.   A Golgi study of
       radial glial cells in developing monkey telencephalon:
       Morphogenesis and transformation into astrocytes.   Anat.
       Embryol.   156: 115-152.

(23)   Skoff, R.P.; Price, D.L.; and Stocks, A.   1976.   Electron
       microscopic autoradiographic studies of gliogenesis in rat
       optic nerve I.   Cell proliferation.   Comp. Neur. 169: 291-311.

(24)   Spray, D.C.; Harris, A.L.; and Bennett, M.V.L.   1981.
       Equilibrium properties of a voltage-dependent junctional
       conductance.   J. Gen. Physiol.   77: 77-93.

(25)   Sternberger, N.H.; Itoyama, Y.; Kies, M.W.; and Webster,
       H. de F. 1978. Myelin basic protein demonstrated immuno-
       cytochemically in oligodendroglia prior to myelin sheath
       formation. Proc. Natl. Acad. Sci. USA 75: 2521-2524.

(26)   Sternberger, N.H.; Quarles, R.H.; Itoyama, Y.; and Webster,
       H. de F. 1979. Myelin-associated glycoprotein demon-
       strated immunocytochemically in myelin and myelin-forming
       cells of developing rat. Proc. Natl. Acad. Sci. USA 76:
       1510-1514.

(27)   Strichartz, G.; Small, R.; Nicholson, C.; Pfenninger, K.H.;
       and Llinas, R. 1980. Ionic mechanisms for impulse propa-
       gation in growing non-myelinated axons: saxitoxin binding
       and electrophysiology. Soc. Neurosci. Abstr. 6: 225.2.

(28)   Sturrock, R.R. 1976. Light microscopic identification of
       immature glial cells in semithin sections of the develop-
       ing mouse corpus callosum. J. Anat. 122: 521-537.

(29)   Sturrock, R.R. 1980. A development study of the mouse
       neostriatum. J. Anat. 130: 243-261.

(30)   Tang, C.M.; Strichartz, G.R.; and Orkand, R.K. 1979.
       Sodium channels in axons and glial cells of the optic
       nerve Necturus Maculosa. J. Gen. Physiol 74: 629-642.

(31)   Vaughan, J.E.; Hinds, P.L.; and Skoff, R.P. 1970. Elec-
       tron microscopic studies of Wallerian degeneration in rat
       optic nerve. I. The multipotential glia. J. Comp. Neurol.
       140: 175-206.

(32)   Webster, H. de F.; Trapp, B.D.; Sternberger, N.H.; and
       Quarles, R.H. 1981. Myelin forming glial cells: morpho-
       logical and immunocytochemical observations. In Sym-
       posium on Development in the Nervous System, eds. D.R.
       Garrod and J.D. Feldman, pp. 265-288. Cambridge University
       Press, in press.

(33)   Wiley, C.A.; Livingston, C.A.; and Ellisman, M.H. 1980.
       Development of axonal membrane specializations defines
       nodes of Ranvier and precedes Schwann cell myelin elabora-
       tion. Dev. Biol. 79: 334-355.

Neuronal-glial Cell Interrelationships, ed. T.A. Sears, pp. 115-130.
Dahlem Konferenzen 1982. Berlin, Heidelberg, New York: Springer-Verlag.

Cellular and Non-Cellular Influences
on Myelin-Forming Cells

R. P. Bunge
Dept. of Anatomy and Neurobiology, Washington University
School of Medicine, St. Louis, MO 63110, USA

Abstract. Both the cause of the breakdown of the oligodendro-
cyte-myelin unit in multiple sclerosis lesions and the reasons
for the failure of repair by remyelination are unknown. In
many experimental demyelinating conditions of the central nervous system substantial remyelination by oligodendrocytes is
observed; in others Schwann cells may provide myelin of the
peripheral type within central neural tissues. The precise environmental requirements for the production of myelin by oligodendrocytes have not been defined. Studies utilizing separated
(and recombined) populations of Schwann cells, sensory neurons,
and fibroblasts in tissue culture indicate that: 1) axonal contact provides a mitogenic signal for Schwann cells, 2) axonal
contact engenders Schwann cell production of certain collagenous
products, 3) secretory activity by Schwann cells appears to be
necessary for the expression of Schwann cell function, and 4)
long-term cultures containing only neurons and Schwann cells
in serum-containing medium will not myelinate unless embryo
extract is added in the culture medium or a population of fibroblasts is present in the culture dish. These observations on
Schwann cells suggest that the expression of oligodendrocyte
function may also have complex, as yet undefined, microenvironmental requirements.

INTRODUCTION

This brief review will consider several of the known influences

on the functional expression of oligodendrocytes and Schwann

cells. These myelin-forming cells have quite different origins -

the oligodendrocyte from the neural tube and the Schwann cell

from neural crest and placode - and they function in the quite

different environments of the central (CNS) and peripheral nervous systems (PNS). Yet they may respond to the same axon (e.g., myelinated axons of certain dorsal root fibers) to provide peripheral myelin along a portion of the axon length and central myelin along the remainder. Because the Schwann cell has been isolated in pure populations in culture and recombined with axons under defined conditions and in isolation from other cell types, rather more can be said about specific requirements for its functional expression. It seems reasonable to suggest, however, that the complex series of cellular and humoral inter- actions which guide the differentiation of the Schwann cell will have certain counterparts in oligodendrocyte development. I will begin by briefly discussing several of the known aspects of the failure of oligodendrocyte function observed in the CNS lesions in multiple sclerosis with an attempt to relate these to known aspects of oligodendrocyte biology. This will be fol- lowed by a more detailed description of certain of the known conditions which may cause Schwann cells to undergo functional failures showing certain similarities to those exhibited by oligodendrocytes in the multiple sclerosis plaque. Finally, similarities and differences in oligodendrocyte and Schwann cell function in the process of myelination will be considered.

FAILURE OF THE OLIGODENDROCYTE-MYELIN UNIT IN MULTIPLE SCLEROSIS

The long-term result of this disease process in certain regions (plaques) of CNS white matter is 1) loss of myelin sheaths and the oligodendrocyte population, 2) retention of axonal continuity, 3) hypertrophy of fibrillar astrocytes, and 4) the presence of substantial extracellular space (3,23,24). The lesion sometimes shows evidence along its borders of cellular activity interpreted as being indicative of an extension of the demyelinated region or, in other cases, activity suggesting that attempts at remyelination of the demyelinated axons occur. Whereas the lesion matrix in its later stages is relatively acellular, it has been established that the perivascular spaces in regions in long standing plaques contain a substantial number of plasma cells, as well as macro- phages. Prineas (24) has emphasized that in regions where myelin deterioration appears to be underway, one can frequently demonstra▸

an association between cells which appear to be phagocytes and myelin sheaths which are in the process of breakdown. Because primary demyelination can occur as a non-specific consequence of cell-mediated immune reactions (36), it has been suggested (7) that macrophages either newly entering or resident within the lesion may secrete proteases which initiate the degradation of myelin. The activation of these macrophages occurs as a part of a cell-mediated immune response occurring to a non-brain antigen in the vicinity of myelinated tissue. Viral products harbored within some cell type within the CNS may provide the antigen to trigger this immune response. This concept presumes that the myelin-oligodendrocyte unit is more susceptible to enzymatic attack than is the axon itself.

Two central questions remain largely unanswered regarding the genesis of the multiple sclerosis plaque. The first is the problem of identifying the earliest changes, and thus presumably identifying the factors which initiate the demyelinating process. This complicated and controversial subject will be considered elsewhere in this volume; several recent reviews which consider the factors which may be involved in the etiology of the lesion are available (3,28,32). The second question is more appropriate to our present discussion: why is the reparative process of re-myelination, which is known to occur in many experimental de-myelinations of CNS tissue, deficient in the repair of the de-mylination in the multiple sclerosis plaque? Whereas some de-gree of remyelinating activity can be observed along the border between the demyelinated zone and the adjacent normal white matter, this is generally limited and does not appear to be suf-ficient to prevent the plaque from expanding. Prineas (25) has recently studied in some detail the degree of remyelination which occurs in multiple sclerosis plaques and has stressed the fact that, whereas there is a general paucity of oligodendro-cytes, the few oligodendrocyte cell bodies present within the lesion may sometimes be seen in direct contact with axons, but are deficient in providing myelin sheaths.

How do conditions within the multiple sclerosis plaque differ
form those (see, for example, (16)) in which oligodendrocyte
remyelination occurs? The conditions in which central remyelination has been observed have been discussed in some detail
by Blakemore (6). In several of the demyelinating conditions
he has observed, especially where the demyelinated areas have
some continuity with the point of attachment of dorsal roots,
Schwann cells can be seen to be involved in the CNS remyelinating process; in other situations they are not. Blakemore
(6) believes that retention of myelin debris inhibits, and the
presence of astrocytes fosters, oligodendrocyte remyelination.
In some CNS demyelinations Schwann cells can be seen to be in
apparent competition with oligodendrocytes in the process of
axonal remyelination. I will discuss below the conditions
which may determine whether oligodendrocytes or Schwann cells
are able to be operative within demyelinated areas of the CNS.
It is clear from these conditions that our inability to define
the conditions under which remyelination in the CNS may or
may not occur is a major deficiency in understanding the patho-
biology of the multiple sclerosis lesion.

SOME KNOWN INFLUENCES ON OLIGODENDROCYTE DEVELOPMENT AND FUNCTION
A variety of experiments have been undertaken in the developing
nervous system which have been aimed at defining those cellular
influences which are instrumental in fostering oligodendrocyte
differentiation. In an extensive study involving the oligodendrocyte population of the optic nerve, Fulcrand and Privat (14)
demonstrated that the presence of axons appears to be essential
for initiation of proliferation and differentiation of the oligodendrocyte population. In these experiments the nerve fibers of
the optic nerve were caused to degenerate by the removal of the
eye, and the effect on populations of glial cells in the optic
nerve was studied at various times after the lesion. Lesions
were produced in rats of different ages and in the analysis of
the results these authors observed that if axons are removed
from contact with the developing oligodendrocytes population
during a period shortly after birth, when oligodendrocyte

differentiation and myelination are in progress, the subsequent
glial population of the optic nerve will contain practically no
oligodendrocytes. Conversely, if the axons are caused to degenerate in older animals, the oligodendrocyte population was subsequently retained for several months within the optic nerve.
These results suggest that oligodendrocyte differentiation is
dependent upon axonal contact during an early stage, but that
after this differentiation has been accomplished, subsequent
contact with axons is not required for oligodendrocyte survival.
In related experiments done in tissue culture, Mirsky et al. (20)
have shown that oligodendrocytes taken from maturing white matter can be placed in tissue culture and retain some of the distinguishing characteristics of oligodendrocytes, particulary the
expression on their surface of the antigen galactocerebroside,
even after many weeks in culture. These observations suggest
that after the oligodendrocyte begins to produce specific products (in this case a myelin-specific glycolipid) this expression
is continued even after the oligodendrocyte is removed from contact with an axon.

These observations during development raise the question of why
the numerous axons retained in the multiple sclerosis lesion do
not signal the oligodendrocytes to mount a more effective remyelination. The question of the sources of oligodendrocytes in
mature CNS tissues which may be available for the process of
myelin repair is not yet settled. It is not clear whether the
repairing oligodendrocytes, when these are able to function
effectively (16), derive from a stem cell population of glio-
blasts or whether they represent proliferation of oligodendrocytes that are already differentiated and have previously been
associated with the process of myelination.

The question must be raised whether the axonal influence discussed above is the sole agency influencing initial oligodendrocyte differentiation. Observations in the mutant mouse jimpy,
where the process of myelination is defective, have suggested
to Skoff (33) that the failure of oligodendrocyte function in

this mutant mouse may be related to a primary deficiency, not
in the oligodendrocyte, but in the astrocyte population. Skoff
has suggested that the development of the astrocyte and its influence on the microenvironment of the axonal and oligodendrocyte population may be critical for the full expression of oligodendrocyte function. It should be noted that Raine (28) has
presented data indicating that the earliest changes in the multiple sclerosis may be in astrocytes.

KNOWN INFLUENCES ON SCHWANN CELL DEVELOPMENT AND FUNCTION
A series of recent tissue culture observations have added considerably to our understanding of the control of Schwann cell
function in the development of the peripheral nerve and in related disease processes; these may be briefly summarized as follows.

Several methods are now available for the preparation of enriched
or purified populations of Schwann cells in tissue culture (19,26,
37). Utilizing one of these methods (37) we have demonstrated
that the growth of sensory neuron axons into a mitotically quiescent population of Schwann cells induces proliferation within
those Schwann cells in direct contact with the growing nerve fibers (38). Schwann cells in the same culture dish which are not
in contact with growing nerve fibers remain quiescent. Thus,
direct contact between axon and Schwann cell appears necessary
for the delivery of this mitogenic signal. Evidence has been
presented that the mitogenic signal from axon to Schwann cell
is located on the axonal surface and may be delivered to the
Schwann cell by particulate fractions prepared from cultured
neurites (29,31), and that signal may be removed from the axolemmal surface by treatment with trypsin (30). These experiments indicate that the number of Schwann cells that are generated to populate the expanding numbers of peripheral nerve fibers during development may be controlled by direct contact
with the expanding axonal surface.

When an adequate number of Schwann cells has been generated to
provide for the ensheathment of unmyelinated nerve fibers and

the myelination of larger nerve fibers, proliferative activity
begins to decline and ensheathment (that is the surrounding of
each of the axons by a portion of Schwann cell cytoplasm) begins (4,34). Recent observations, again in tissue culture preparations, have established that in a subsequent step in their
development, Schwann cells have the capability, when in contact
with axons, of providing the basal lamina which normally is found
on their surface, as well as a small number of collagenous fibrils in relation to this basal lamina (10). This was established by electron microscopic study of tissue culture preparations in which Schwann cells were allowed to grow in relation
to axons in the absence of a fibroblast population. Therefore,
these connective tissue components could be reliably established
as being provided by the Schwann cell population and not by
fibroblasts. More recent observations (21) indicate that in the
presence of defined tissue culture medium (8) (lacking the usual
components of serum and embryo extract), Schwann cell proliferation in relation to axonal growth occurs, but that the subsequent steps of ensheathment of axons and the myelination of the
larger axons are not observed even after many weeks in culture.
Whereas a plethora of Schwann cells is generated in relation
to the axons in cultures in defined medium, these Schwann cells
do not progress to enclose axons (the process of ensheathment)
or to provide myelin sheaths. Furthermore, they have lost the
ability to secrete visible basal lamina or collagenous fibrillar
material. This arrest in Schwann cell development and secretion
is rapidly reversed, however, if serum and embryo extract are
added to the defined medium; within one week after the addition
of these undefined components, myelin segments begin to appear
in the culture dish. Further experiments have recently shown
that, of the two undefined components added, the more critical
component is embryo extract. In the absence of fibroblasts,
Schwann cells related to axons will not produce myelin in the
presence of serum alone. These experiments indicate that there
is present in embryo extract some essential component for the
furtherance of Schwann cell function; they further suggest that
in the absence of Schwann cell secretion normal ensheathment of

axons by the Schwann cell cannot occur. Additional support of
this point of view is provided by the observation that the use
of a proline analogue to disrupt collagen synthesis by Schwann
cells interferes with their ability to provide axonal ensheathment (12).

Additional experiments (again in tissue culture) indicate that
the Schwann cell in relation to the axon is not able to complete
its ensheathment function if it does not have simultaneous contact with a second surface component, i.e., a component in addition to the axons itself (11). These observations derive from
cultures in which Schwann cells are growing in relation to axons
that are not in concomitant contact with the usual tissue culture substrate (for this kind of preparation, a reconstituted
collagen film). This circumstance is observed when axons course
from bulbous explant regions to make contact with the tissue
culture substrate some distance from the explants. In the inter-
vening span these axons are surrounded only by tissue culture
medium and are not in contact with the underlying substratum.
Schwann cells, under these circumstances, can be seen to relate
to the bundles of axons that are suspended in tissue culture
medium, but the Schwann cells are unable to expand their population, to spread over the surface of the axons, and to go on to
ensheathe or myelinate the axonal components. This observation
has led to the suggestion that not only is contact with an axon and
secretion of extracellular matrix components necessary for full
Schwann cell function, but also simultaneous contact with a
"third element" which provides a necessary substrate for Schwann
cell expansion and proper contact with the axonal shaft.

These observations on Schwann cells indicate the degree of complexity that may be involved in obtaining full performance of
the myelin-forming cell of the PNS. They establish that contact
with an axon is needed to adjust the number of Schwann cells
present. It is also known from work in vivo that the axon will
provide a signal to the Schwann cell which will determine whether
the Schwann cell will be an ensheathing Schwann cell for an

unmyelinating nerve fiber or a myelinating Schwann cell for a myelinated nerve fiber (1,35). In order to accomplish full differentiation, however, the Schwann cell appears to be required to undertake an intermediate secretory phase in which it produces certain extracellular matrix components, without which it cannot obtain the correct relationship to the axon and will become arrested at an early stage in its development. There appears to be an additional requirement for the Schwann cell to contact an extracellular matrix component in order to complete its differentiation. This latter observation could explain the positions in the body where Schwann cells are able to function normally. Under normal conditions Schwann cells do not invade the CNS, nor are they able to invade certain parts of the periphery which do not contain the apparently necessary extracellular matrix components. An example is provided by the epidermis of the skin, where nerve endings end among the epidermal cells without accompanying Schwann cell ensheathment.

The discussion in the immediately preceding paragraphs applies to Schwann cells in the process of normal development (34). The reactions of mature Schwann cells to injury are less well understood (4). It can be noted that in general peripheral nerve responds to injury with a substantial degree of remyelination. In contrast to the CNS, particularly the multiple sclerosis plaque as discussed above, there are few circumstances in which a paucity of Schwann cells appears to contribute to the failure of remyelination subsequent to a peripheral demyelinating episode.

There is, however, one instance in which Schwann cell development is seriously lacking and the axons of the PNS can be found to be entirely free of Schwann cell ensheathment and lying in direct contact with one another. This condition is found in the PNS of several strains of the dystrophic mouse, particularly in both dorsal and ventral nerve roots (9). In these regions it has been established that Schwann cell function is seriously deficient from the time of initial myelination throughout adulthood. In regions of the nerve roots, large numbers of axons

(some being axons of considerable diameter and certainly competent to accept a myelin sheath) are seen to be lying in direct
contact with one another, not separated, as normally, by inter-
vening Schwann cell processes and the extracellular matrix components of the endoneurium. In these regions there are small
numbers of cells which appear to be undifferentiated Schwann
cells. They lie adjacent to bundles of naked axons and exhibit
an occasional remnant of basal lamina covering. They appear
unable to ensheathe or myelinate the adjacent axons. It should
be noted that more subtle abnormalities are present in other
regions of the PNS as well (17).

We have presented elsewhere (22) our arguments that the deficiency in these nerve roots may be explained not as a failure
of axonal or Schwann cell function, but on some failure in the
generation of the extracellular matrix components which (as we
have emphasized above) must be present for the full expression
of Schwann cell differentiation. The point of this brief discussion regarding the Schwann cell is to emphasize that myelin-
forming cells may fail in their functions, not because they
themselves are deficient, but because they find themselves in
an environment that is not conducive to normal function.

SOME COMPARISONS BETWEEN THE FUNCTIONAL ACTIVITY OF SCHWANN CELLS AND OLIGODENDROCYTES

Areas of CNS demyelination may be induced by a number of methods
which allow the survival of axons and subsequent remyelination.
Recently a number of experiments have been reported in which
Schwann cells are deliberately introduced into these regions
to observe the degree of CNS remyelination provided by Schwann
cells. In the experiments undertaken by Blakemore (5), a segment of complete peripheral nerve was introduced into an area
of demyelination in which oligodendrocyte proliferation in response to the lesion was inhibited by the use of irradiation.
Myelination by both oligodendrocytes and Schwann cells was observed; Schwann cell myelination occurred primarily along the
margins of the implanted nerve. Duncan et al. (13) have placed

pure populations of Schwann cells (along with a fragment of
collagen substratum) into demyelinated regions of the dorsal
columns of the mouse spinal cord and observed PNS myelination
in the region directly adjacent to the implant. The limited
area of repair observed in both instances raises the question
of whether microenvironmental conditions are only sufficient
for Schwann cell function if some connective tissue scarring
attendent to the surgery has occurred. This scarring with fibroblast proliferation could provide the "third element" of
extracellular matrix material which the tissue culture experiments discussed above indicate is necessary for full Schwann
cell differentiation. When Schwann cell invasion is induced to
occur in the neonatal CNS by the application of radiation, the
regions occupied by Schwann cells are accompanied by readily
demonstrable connective tissue components (15). Whether these
are products of the invading Schwann cells or of co-migrating
fibroblasts has not been resolved. The converse experiment,
where segments of CNS tissue (optic nerve) are placed into the
PNS (2), demonstrates that oligodendrocytes are able to provide
myelin around the very few PNS axons which are able to penetrate
a CNS graft in this position.

Recently techniques have been described for growing separate
populations of astrocytes and oligodendrocytes in culture (18)
and culturing sensory neurons (free of Schwann cells) which
provide axons capable of inducing myelin formation by added
CNS glial cells (39,40). This latter preparation provides a
method of directly assaying oligodendrocyte myelin production
in dissociated cell cultures (Fig. 1a,b). Experiments can now
be designed to define the microenvironmental requirements for
oligodendrocyte function in tissue culture and the identification of inhibitory conditions. This may lead to a better understanding of the oligodendrocyte functional failure in the multiple sclerosis plaque.

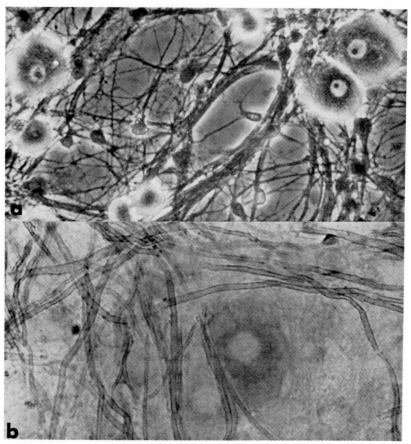

FIG. 1 - These light micrographs illustrate dorsal root ganglion
neurons dissociated from 19-21 day rat embryos treated for the
initial 2 wks in culture with media containing fluorodeoxyuridine
(FUDR) to remove the indigenous Schwann cells. After 1 wk in
media without FUDR, mixed glial populations were added from the
spinal cords of 14-17 day rat embryos. In (a) a living culture
is shown by phase microscopy 16 days after receiving dissociated
glial cells from the ventral commisure region of the spinal cord
of a 17 day rat embryo. The glial population will be substantially expanded prior to the onset of myelination. In (b) a
similar culture is shown 4 wks after the addition of glia and
subsequent to Sudan black staining. The background contains
large sensory neurons; the many nerve fibers have now been rich-
ly populated with glial cells which have formed myelin sheaths
typical of those found within the central nervous system. From
cultures prepared and photographed by Dr. Patrick Wood. (a) X 390,
(b) X 585.

Acknowledgements. Work in the author's laboratory is supported by grants NSO9923 and GM 28002 from the National Institutes of Health and Grant RG 1118 from the National Multiple Sclerosis Society.

REFERENCES

(1) Aguayo, A.J.; Charron, L.; and Bray, G.M. 1976a. Potential of Schwann cells from unmyelinated nerves to produce myelin: a quantitative ultrastructural and radiographic study. J. Neurocytol. 5: 565-573.

(2) Aguayo, A.J.; Dickson, R.; Trecarten, J.; Attiwell, M.; Bray, G.M.; and Richardson, P. 1978. Ensheathment and myelination of regenerating PNS fibres by transplanted optic nerve glia. Neurosci. Lett. 9: 97-104.

(3) Andrews, J.M. 1972. The ultrastructural neuropathy of multiple sclerosis. In Multiple Sclerosis, Immunology, Virology, and Ultrastructure, eds. F. Wolfgram, G.W. Ellison, J.G. Stevens, and J.M. Andrews. New York: Academic Press.

(4) Asbury, A.K. 1975. The biology of Schwann cells. In Peripheral Neuropathy, eds. P.J. Dyck, P.K. Thomas, and E.H. Lambert, pp. 201-212. Philadelphia: W.B. Saunders.

(5) Blakemore, W.F. 1977. Remyelination of CNS axons by Schwann cells transplanted from the sciatic nerve. Nature 266: 68-69.

(6) Blakemore, W.F. 1978. Observation on remyelination in rabbit spinal cord following demyelination induced by lysolecithin. Neuropathol. & Appl. Neurobiol. 4: 47-59.

(7) Bloom, B.R.; Ju, G.; Brosnan, C.; Cammer, W.; and Norton, W. 1978. Notes on the pathogenesis of multiple sclerosis. Neurology 28: 93-101.

(8) Bottenstein, J.E., and Sato, G.H. 1979. Growth of a rat neuroblastoma cell line in serum-free supplemented medium. Proc. Natl. Acad. Sci. (USA) 76: 514-517.

(9) Bradley, W.B., and Jenkison, M. 1973. Abnormalities of peripheral nerve in murine muscular dystrophy. J. Neurol. Sci. 18: 227-247.

(10) Bunge, M.B.; Williams, A.K.; Wood, P.M.; Uitto, J.; and Jeffrey, J.J. 1980. Comparison of nerve cell and nerve cell plus Schwann cell cultures, with particular emphasis of basal lamina and collagen formation. J. Cell Biol. 84: 184-202.

(11) Bunge, R.P., and Bunge, M.B. 1978. Evidence that contact with connective tissue matrix is required for normal interaction between Schwann cells and nerve fibers. J. Cell Biol. 78: 943-950.

128 R.P. Bunge

(12) Copio, D.S., and Bunge, M.B. 1980. Use of a proline analogue to disrupt collagen synthesis prevents normal Schwann
cell differentiation. J. Cell Biol. 87: 114a.

(13) Duncan, I.D.; Aguayo, A.J.; Bunge, R.P.; and Wood, P. 1979.
Transplantation of rat Schwann cells cultured in vitro into
demyelinated areas of the mouse spinal cord. Soc. Neurosci.
Abstr. 5: 510.

(14) Fulcrand, J., and Privat, A. 1977. Neuroglial reactions
secondary to wallerian degeneration in the optic nerve of
the postnatal rat: Ultrastructural and quantitative study.
J. Comp. Neurol. 176: 189-224.

(15) Gilmore, S.A., and Duncan, D. 1968. On the presence of
peripheral-like nervous and connective tissue within ir-
radiated spinal cord. Anat. Rec. 160: 675-690.

(16) Herndon, R.M.; Price, D.L.; and Weiner, L.P. 1977. Regeneration of oligodendria during recovery from demyelinating
disease. Science 195: 693-694.

(17) Jaros, E., and Bradley, W.G. 1979. Atypical axon-Schwann
cell relationships in the common peroneal nerve of the dystrophic mouse: an ultrastructural study. Neuropathol. &
Appl. Neurobiol. 5: 33-147.

(18) McCarthy, K.D., and deVellis, J. 1980. Preparation of
separate astroglial and oligodendroglial cell cultures
from rat cerebral tissue. J. Cell Biol. 85: 890-902.

(19) McCarthy, K.D., and Partlow, L.M. 1976. Neuronal stimulation of 3H-thymidine incorporation by primary cultures of
highly purified non-neuronal cells. Brain Res. 114: 415-426.

(20) Mirsky, R.; Winter, J.; Abney, E.R.; Pruss, R.M.; Gavrilovic,
J.; and Raff, M.C. 1980. Myelin-specific proteins and glycolipids in rat Schwann cells and oligodendrocytes in culture. J. Cell Biol. 84: 483-494.

(21) Moya, F.; Bunge, M.B.; and Bunge, R.P. 1980. Schwann cells
proliferate but fail to differentiate in defined medium.
Proc. Natl. Acad. Sci. (USA) 77: 6902-6906.

(22) Okada, E.; Bunge, R.P.; and Bunge, M.B. 1980. Abnormalities expressed in long term cultures of dorsal root ganglia
from the dystrophic mouse. Brain Res. 194: 455-470.

(23) Prineas, J. 1975. Pathology of the early lesion in multiple sclerosis. Hum. Pathol. 6: 531-554.

(24) Prineas, J.W., and Connell, F. 1978. The fine structure
of chronically active multiple sclerosis plaques. Neurol.
28: 68-75.

(25) Prineas, J.W., and Connell, F. 1979. Remyelination in multiple sclerosis. Ann. Neurol. 5: 22-31.

(26) Raff, M.; Abney, E.; Brockes, J.; and Hornby-Smith. A. 1978. Schwann cell growth factors. Cell 15: 813-822.

(27) Raine, C.S. 1976. On the occurrence of Schwann cells within the normal central nervous system. J. Neurocytol. 5: 371-380.

(28) Raine, C.S. 1977. The etiology and pathogenesis of multiple sclerosis: recent developments. Pathobiol. Annual 7: 347-384.

(29) Salzer, J.L., Bunge, R.P. 1980. Studies of Schwann cell proliferation: I. An analysis in tissue culture of proliferation during development, Wallerian degeneration, and direct injury. J. Cell Biol. 84: 739-752.

(30) Salzer, J.L.; Bunge, R.P.; and Glaser, L. 1980a. Studies of Schwann cell proliferation: III. Evidence for the surface localization of the neurite mitogen. J. Cell Biol. 84: 767-778.

(31) Salzer, J.L.; Williams, A.K.; Glaser, L.; and Bunge, R.P. 1980b. Studies of Schwann cell proliferation: II. Characterization of the stimulation and specificity of the response to a neurite membrane fraction. J. Cell Biol. 84: 753-766.

(32) Seil, F.J. 1977. Tissue culture studies of demyelinating disease: a critical review. Ann. Neurol. 2: 345-355.

(33) Skoff, P. 1976. Myelin deficit in the Jimpy mouse may be due to cellular abnormalities in astroglia. Nature 264: 560-562.

(34) Webster, H. deF. 1975. Development of peripheral myelinated and unmyelinated nerve fibers. In Peripheral Neuropathy, eds. P.J. Dyck, P.K. Thomas, and E.H. Lambert, pp. 37-61. Philadelphia: Saunders.

(35) Weinberg, H., and Spencer, P. 1975. Studies of myelinogenesis, I. Myelination of regenerating axons after entry into a foreign unmyelinated nerve. J. Neurocytol. 4: 395-418.

(36) Wisniewski, H.M., and Bloom, B.R. 1975. Primary demyelination as a nonspecific consequence of a cell-mediated immune reaction. J. Exp. Med. 141: 346-359.

(37) Wood, P.M. 1976. Separation of functional Schwann cells and neurons from normal peripheral nerve tissue. Brain Res. 115: 361-375.

(38) Wood, P.M., and Bunge, R.P. 1975. Evidence that sensory axons are mitogenic for Schwann cells. Nature 265: 662-664.

130

(39) Wood, P.M., and Bunge, R. 1980. Dorsal root ganglion neurons stimulate differentiation and myelin formation by oligodendrocytes derived from embryonic spinal cord. Abstr. Soc. Neurosci. 6: 379.

(40) Wood, P.; Okada, E.; and Bunge, R. 1980. The use of networks of dissociated rat dorsal root ganglion neurons to induce myelination by oligodendrocytes in culture. Brain Res. 196: 247-252.

Neuronal-glial Cell Interrelationships, ed. T.A. Sears, pp. 131-146.
Dahlem Konferenzen 1982. Berlin, Heidelberg, New York: Springer-Verlag.

Cell Interactions and the Cytoskeleton

F. Solomon
Dept. of Biology and Center for Cancer Research
Massachusetts Institute of Technology, Cambridge, MA 02138, USA

Abstract. The cytoskeleton is traditionally identified as the
organelle of cellular motility. A long list of experiments
suggests that that function is in part performed by a transmembrane connection, linking the cell surface and the cytoskeleton. The structural and molecular bases for that connection are not known. Recent studies of the cytoskeleton itself, concentrating on in situ analysis, have provided new
approaches to this problem. They have altered our view of
cytoskeletal organization and expanded the possible functions of the transmembrane connection.

INTRODUCTION

The importance of specific cell interactions in differentiated cell function is well-established. Their role in the development and activities of the nervous system is described
in the proceedings of a previous Dahlem Conference (48), and
it is the subject of other contributions to this meeting. The
goal of this paper is to discuss one level upon which these
phenomena occur: the transduction of information between the
external surface of the cell and the internal cytoskeleton.
In particular, it will focus on new views of the cytoskeleton
itself which alter our view of its organization and function.

The notion of a "transmembrane connection" has acquired the
status of inevitability. It presents an image of cytoskeletal
elements - microtubules, microfilaments, or intermediate

filaments - in direct or indirect contact with species which span the membrane and which themselves are in contact, direct or indirect, with the extracellular environment. There is, however, almost no information on the structural or molecular basis for such a connection. With the exception of one model system, we have not identified the crucial transmembrane components, nor do we know the fine structure of the cytoskeletal elements near the membrane. Instead, our knowledge of the transmembrane connection is largely inferred from an impressive accumulation of phenomena. Those results have been reviewed extensively by others (34, 45, 49, 50). Briefly, the most important lines of evidence are:

1. The motile functions of the cell are known to require the cytoskeleton. The vectorial components of those functions - the precise shape that a spreading cell assumes (16), the direction of cell movement (29), or the direction and termination of axonal elongation (54, 59) - can all be specified by extracellular signals.

2. Specific cell surface components are distributed non-randomly in the plane of the membrane - maintaining that microheterogeneity depends upon an intact cytoskeleton, and does not survive disruption of microtubules (51).

3. The mobility of surface components is affected by the state of the cytoskeleton (23).

4. The distribution of some surface components is partially coincidental with the distribution of the microfilaments of the same cell (1, 3, 35, 56). In other cases, such as the acetylcholine receptors of myotubes, the patterns may instead be perfectly complementary (9).

5. The effect of extracellular ligands such as mitogens is modulated by the state of microtubule assembly (47, 52).

6. The molecular species which link actin to a transmembrane protein in the human erythrocyte have been identified (5, 6). The human erythrocyte is an excellent model system because of the unique simplicity of its membrane and its cytoskeleton, although that simplicity is likely to make any extrapolation to other cells problematical.

This somewhat ruthless abstract of an enormous body of work identifies the primary support for existence of a transmembrane connection. Clearly, the solutions to many of the questions posed by these results require identification of the structural and molecular species involved at the membrane. That problem is under active investigation in several laboratories.

The internal component of the transmembrane connection - the cytoskeleton - has been the object of intensive structural and molecular research for almost thirty years. Electron microscopy has established the ultrastructure of each of the cytoplasmic fibers. Immunofluorescence has helped visualize the distribution of each in the cell. In vitro systems have been devised for analyzing molecular components and kinetics of assembly of microtubules and microfilaments. Finally, drugs which disrupt the cytoskeleton have been used to probe its functions. The basic facts established by this work are the subjects of recent reviews (17, 21, 41).

Recently, new approaches to the cytoskeleton have provided us with different sorts of information. For example, there is increasing evidence of interactions among the elements of the cytoskeleton, evidence that alters our view of individual filaments with unique functions. Second, analysis of cytoskeletal structures has identified special features at the ends of fibers, which may take us close to the membrane. Finally, work on regulation of macromolecular metabolism indicates that the interactions between the cell membrane, the outside environment and the internal skeleton may play an important role in growth control. These three areas, described in detail below, have altered our view of the organization, structure, and function of the cytoskeleton.

ORGANIZATION OF CYTOSKELETON COMPONENTS
An interaction between the elements of the cytoskeleton is implicit in the simplest experiments on cell motility. Recent

results have begun to characterize those interactions in detail.

The Effect of Microtubule Depolymerizing Drugs
The standard assay for determining if microtubles are involved
in any particular cell process is to add a microtubule-
depolymerizing drug, like colchicine, and watch for inhibition
of the process. It is from that sort of experiment that the
correlation is made between microtubule assembly and maintenance of cellular asymmetry, as well as many other putative
microtubule functions. The biochemical aspects of that para-
digm are probably correct, since no high-affinity colchicine
binding species has been found other than the major microtubule component, tubulin. The structural aspects of that experiment are not as unambiguous. The depolymerization of all
cytoplasmic microtubules is certain to affect other cytoplasmic structures. In particular, in several cell lines, including neuronal cells, microtubule depolymerization is followed
by a complete rearrangement of intermediate filaments (30, 36-
38). Typically, they condense to form tight coils around or
near the nucleus.

This observation not only clouds the traditional interpretation of several experiments, it also raises the possibility
that the microtubule and intermediate filament network are
structurally related to one another. At least at the level of
resolution of the light microscope, the distributions of the
two fibers are largely, but not exclusively, coincident with
one another (28). It is not known if the distribution of
either one is specified by the other, or if there are physical
connections between them.

Structure of the Cytoplasm
Further evidence for microtubule-intermediate filament interactions and other interactions comes from studies of cytoplasmic flow. It is well-known that the cytoplasm of the neuron is strikingly compartmentalized. The cellular synthetic
machinery is confined to the soma, and a variety of structures
and macromolecules are transported down the axon.

Lasek and co-workers have analyzed the molecular aspects of
one such system and have arrived at important conclusions
about structure (8, 33). They identify five classes of an-
terograde transport down the axon. Each class has a unique
velocity; the fastest differs from the slowest by more than
two orders of magnitude. In addition, each class contains its
own unique set of molecular components. Even during the several days required to travel the length of the axon, the components of one class which began together do not diffuse
apart. One coherent class - the slowest moving - contains
tubulin and the intermediate filament proteins. A major component of the next class is actin. In addition, other proteins normally thought of as "soluble" - such as glycolytic
enzymes - are also associated with a discrete, coherently
transported class (22, 39).

At present, little is known about the state of these macromolecules as they are transported. There is evidence that the
majority of the tubulin and actin are in filamentous form,
and intact cytoskeletal fibers are visualized throughout the
axon (39). Such an "assembled" form of the fibers, along with
other proteins, could form a discrete associated structure
which can be distinguished as a unit for transport. The
mechanism of movement remains, at the least, perplexing.
Models for monotonic intracellular transport have been proposed in other cell systems such as the fish erythrophore,
involving a calcium-mediated action of actin-myosin. The
rates of transport in those systems are about the same as
the fastest component of axoplasmic transport, and models of
actin-myosin involvement have been proposed (24, 44). Ac-
commodating the slower rates will require either identification of different "motors" or unique structural features of
the association between the actin-myosin system and each of
the classes of transport.

The Microtrabeculae
Evidence described above directly supports the notion of a

connection between microtubules and intermediate filaments.
The association of microtubules with microfilaments is suggested by the axonal transport data and is implicit in even
the simplist assignments of function as the "bones" and "muscle" of the cell. Direct evidence for such an association
is presented by Porter and his colleagues, who have used a
high voltage electron beam to penetrate parts of fixed but
unsectioned cells. Three-dimensional viewing of these images
reveals that the flocculent "ground substance" evident in thin
sections form a lattice in which the three cytoplasmic fibers
are interconnected by shorter fibers of variable diameter (15,
60). In addition, suspended in these "microtrabeculae" are
other organelles, vesicles, and clusters of ribosomes. The
possible functional significance of these associations is described below. The cellular distribution and fine structure
of the network are dramatically altered during changes in cell
shape or concerted intracellular transport.

The evidence for a microtrabecular network is not universally
accepted. For example, elegant images of cells which are
detergent-extracted before fixation show no such connecting fibers (32). That technique could, of course, produce different
artifacts. In one sense, the microtrabeculae may represent a
static image of a set of connections which rapidly form and
break, and that image may be misleading simply because of its
solidity.

MOLECULAR COMPONENTS OF THE CYTOSKELETON
The major protein components of each of the skeletal fibers
can be induced to assemble into smooth linear polymers. It is
obvious that the dimensions of those polymers make impossible
a direct connection between them and some membrane polypeptide; they are just too big (43). One possibility is that
some associated species are necessary which can bridge this
difference in scale.

Microtubule-associated proteins have been identified primarily
by in vitro analysis. Current evidence suggests that all of

those proteins examined to date are present along the entire
length of the fibers, and that none are localized uniquely at
the distal ends (10, 18, 19, 55). Recent results from in situ
analyses óf microtubules suggest that those associated species
are specific for different structural and functional contexts
(20, 58, 61), but none have been demonstrated to interact with
membrane.

The approaches to the same problem have been more successful
in the case of microfilaments. There are clearly non-actin
proteins distributed along the length of the microfilament
(53). In addition, attention has focused on two specialized
structures which should contain microfilament-membrane interactions. The first is the attachment plaque, regions where
the cell membrane comes into close contact with the surface.
Such regions can be identified in the electron microscope as
points where microfilaments terminate, or begin (31). This
junction of cytoskeleton, membrane, and extracellular protein
is also the unique locale of a 130K dalton polypeptide, called
vinculin (27). Microinjected vinculin also ends up in these
attachment plaques (14) so it can be on the internal surface
of the membrane. All the available data, however, suggest
that it is not a transmembrane protein, so there must be at
least another link in the connection.

The second structure of interest is the brush border of the
intestinal epithelium. Each cell in this organ extends about
one thousand microvilli and each microvillus contains a microfilament cytoskeleton. Two polypeptides of interest have been
found associated with those filaments. One is villin (11),
which is part of the core structure. Villin binds to both monomeric and polymerized actin in vitro (13), but to date has
only been found in brush border. A second protein found in
microvilli, fimbrin (12) is also detected in the surface ruf-
fles of cultured cells, but not along the length of the microfilaments. A protein of that description is a candidate for
one component of the transmembrane connection.

REGULATION OF CELL GROWTH BY THE STATE OF THE CYTOSKELETON
It is well-known that normal cells, established in culture,
will thrive only if allowed to attach to a suitable substratum. If prevented from spreading, they will first stop proliferating and eventually die (46). On the other hand, the
growth of transformed cells is not restricted by "anchorage
dependence" - they can grow without attachment. In 1975,
Folkman and Greenspan examined this and related phenomena in
detail and suggested that the essential element for growth
control in normal cells was the shape of the cell, which was
altered by anchorage to a solid substratum (25). There are a
number of results consistent with this notion, such as sub-
strate-induced carcinogenesis. Direct tests of this hypothesis now have been provided by the work of Folkman and of Penman and their colleagues. Their experiments have refined our
ideas of anchorage dependence and implicated the cytoskeleton
as transducer of signal between membrane and nucleus.

Demonstration of the relationship between growth and shape is
provided by studies of cultured cells (26). Cells are plated
on an adhesive substratum which has been coated with various
thicknesses of a non-adhesive polymer. Depending upon the
thickness of the polymer, the cells spread to varying extents.
The cells which are least well spread show lower rates of DNA
synthesis and a lower index of labelled nuclei. As the polymer is adjusted to induce greater spreading, the level of DNA
synthesis increases in a smooth function. Further results
suggest that the relationship between spreading and DNA synthesis is responsible for the diminished growth in confluent
cultures which previously had been ascribed to "contact inhibition." Crowded cells on a normal substratum are rounder
than sparse cells on the same substratum. When sparse cells
are restricted to the same shape as crowded ones, by plating
on polymer, their rate of DNA synthesis is like that of the
crowded cells.

The implication of those experiments is that cellular geometry
can specify cellular metabolism. The underlying molecular
events in that relationship have now been approached in a
series of experiments. The basis of that work has been the
use of cells gently extracted with non-ionic detergents so
that all the lipids and most of the proteins are released.
What remain are the primarily skeletal fibers. Attached to
that skeleton are most of the polysomal ribosomes as detected
by both microscopic and biochemical techniques (42). A similar association between polysomes and the microtrabeculae is
visualized in unextracted cells using the high voltage electron microscope (24). There are, of course, many possibilities for artifactual associations. Support for the in vivo
significance of the polysome-cytoskeleton association is provided by experiments with polio virus. When HeLa cells are
infected with polio virus, synthesis of host proteins is shut
off and only viral proteins are made. In parallel, host cell
message, normally attached to the skeleton, is displaced in
infected cells by polysomes containing viral messenger RNA.
So the association of active message and the cytoskeleton may
indeed play an in vivo role and connect for the first time
the state of the cytoskeleton and macromolecular metabolism.

This approach has been used to analyze the molecular events of
anchorage dependence (4, 7). In cells deprived of their
normal attachment, the rate of protein synthesis drops gradually to less than 20 percent of the normal level in 72 hours.
The synthesis of new mRNA drops more rapidly - to 20% in 12
hours - but the rate of breakdown also declines so that the
total amount of messenger RNA is the same. Detergent extraction of the suspended cells shows that most of the ribosomes
are no longer attached to the skeleton. Upon reattachment,
the rate of protein synthesis increases rapidly - six to
sevenfold in four hours - without a comparable increase in the
rate of mRNA synthesis, and the polyribosomes are again attached to the skeleton. The restoration of protein synthesis,
and of polysome attachment, require only contact with the

substratum. If cells are allowed to attach but are prevented
from spreading either by drugs or by plating on appropriate
thicknesses of non-adhesive polymer, both of these events
are induced. But DNA synthesis does not resume until the
cells are permitted to spread by removing the drug or plating
on a more adhesive substratum. The nature of this differential control is not known, but it suggests a discrete role for
the cytoskeleton in modulating extracellular influences on
macromolecular metabolism.

CONCLUSION
The discussion above presents recent results from new approaches
to the structure and function of the cytoskeleton. Those
approaches have emphasized analysis of cytoskeletal elements
in situ rather than in the in vitro systems which previously
have been exploited so extensively. From several lines of
evidence using disparate methods - ultrastructural, molecular,
and phenomenological - the new view which is emerging is of a
cytoskeleton intrinsically connected both among its elements
and with other components of the cell. The cytoskeletal role
as transducer of signals from cellular interactions is expanded. Among the issues which remain are elucidation of the detailed molecular basis for these connections and, ultimately,
identification of the fundamental organizing principles
which specify the role of the cytoskeleton. That question has
previously been examined in another context - the determination of patterns of cell shape (2, 40, 57). It is possible
that the scope of those discussions can now be broadened to
include other elements of cellular function.

Cell Interactions and the Cytoskeleton   141

REFERENCES

(1) Albertini, D.F., and Clark, J.I. 1975. Membrane-micro-
tubule interactions: Conconavalin A capping induced redistribution of cytoplasmic microtubules and colchicine
binding proteins. Proc. Nat. Acad. Sci. 72: 4976-4980.

(2) Albrecht-Bühler, G. 1977. Phagokinetic tracks of 3T3
cells: parallels between the orientation of track segments and of cellular structures which contain actin or
tubulin. Cell 12: 333-339.

(3) Ash, J.F.; Louvard, D.; and Singer, S.J. 1977. Antibody-
induced linkages of plasma membrane proteins to intracellular actomyosin-containing filaments in cultured fibroblasts. Proc. Nat. Acad. Sci. 74: 5584-5588.

(4) Benecke, B.-J.; Ben-Ze'ev, A.; and Penman, S. 1978. The
control m-RNA production, translation and turnover in suspended and reattached anchorage-dependent fibroblasts.
Cell 14: 931-939.

(5) Bennett, V., and Stenbuck, P. 1979. Identification and
partial purification of ankyrin, the high affinity membrane attachment site for human erythrocyte spectrin. J.
Biol. Chem. 254: 2533-2541.

(6) Bennett, V., and Stenbuck, P. 1979. The membrane attachment protein for spectrin is associated with band 3 in
human erythrocyte membranes. Nature 280: 468-473.

(7) Ben-Ze'ev, A.; Farmer, S.R.; and Penman, S. 1980. Protein synthesis requires cell-surface contact while nuclear events respond to cell shape in anchorage-dependent
fibroblasts. Cell 21: 365-372.

(8) Black, M., and Lasek, R.J. 1980 Slow components of axonal transport: two cytoskeletal networks. J. Cell Biol.
86: 616-623.

(9) Bloch, R.J., and Geiger, B. 1980. The localization of
acetylcholine receptor clusters in areas of cell-substrate
contact in cultures of rat myotubes. Cell 21: 25-36.

(10) Bolinski, J.C., and Borisy, G.G. 1980. Immunofluorescence localization of HeLa cell maps on microtubules in
vitro and in vivo. J. Cell Biol, in press.

(11) Bretscher, A., and Weber, K. 1979. Villin: the major
microfilament-associated protein of the intestinal microvillus. Proc. Nat. Acad. Sci. 75: 2321-2325.

(12) Bretscher, A., and Weber, K. 1980. Fimbrin, a new
microfilament-associated protein present in microvilli
and other cell surface structures. J. Cell Biol. 86:
335-340.

(13) Bretscher, A., and Weber, K. 1980. Villin is a major protein of the microvillus cytoskeleton which binds both F and F actin in a calcium-dependent manner. Cell 20: 839-847.

(14) Burridge, K., and Feramisco, J. 1980. Microinjection and localization of a 130K protein in living fibroblasts: a relationship to actin and fibronectin. Cell 19: 587-595.

(15) Byers, H.R., and Porter, K.R. 1977. Transformation in the structure of the cytoplasmic ground substance in erythrophores during pigment aggregation and dispersion. J. Cell Biol. 75: 541-558.

(16) Carter, S.B. 1967. Haptotoxis and the mechanism of cell motility. Nature 213: 256-267.

(17) Clarke, M., and Spudich, J. 1977. Nonmuscle contractile proteins. Ann. Rev. Biochem. 46: 797-822.

(18) Connolly, J.A.; Kalnins, V.I.; Cleveland, D.W.; and Kirschner, M.W. 1977. Immunofluorescent staining of cytoplasm and spindle microtubules in mouse fibroblasts with antibody to tau protein. Proc. Nat. Acad. Sci. 74: 2437-2440.

(19) Connolly, J.A.; Kalnins, V.I.; Cleveland, D.W.; and Kirschner, M.W. 1978. Intercellular localization of the high molecular weight microtubule accessory protein by immunofluorescence. J. Cell Biol. 76: 781-786.

(20) Duerr, A.; Pallas, D.; and Solomon, F. 1980. Molecular analyses of cytoplasmic microtubules. J. Cell Biol., in press.

(21) Dustin, P. 1978. Microtubules. Berlin: Springer Verlag.

(22) Eckert, B.S.; Koons, S.J.; Schontz, A.W.; and Zokel, C.R. 1980. Association of creatine phospholinase with the cytoskeleton of cultured mammalian cells. J. Cell Biol. 86: 1-5.

(23) Edelman, G.M. 1976. Surface modulation in cell recognition and growth. Science 192: 218-226.

(24) Ellisman, M. 1981. Beyond microtubules and microfilaments. In Cytoarchitecture of the Nervous System, eds. R. Lasek and M. Shelanski. Neurosciences Research Program, in press.

(25) Folkman, J., and Greenspan, H. 1975. Influence of geometry on control of cell growth. Biochim. Biophys. Acta 417: 211-236.

Cell Interactions and the Cytoskeleton 143

(26) Folkman, J., and Moscona, A. 1978. Role of cell shape in growth control. Nature 273: 345-349.

(27) Geiger, B. 1979. A 130K protein from chicken gizzard: its localization at the termination of microfilament bundles in cultured chicken cells. Cell 18: 193-205.

(28) Geiger, B., and Singer, J. 1980. Association of microtubules and intermediate filaments in chicken gizzard cells as detected by double immunofluorescence. Proc. Nat. Acad. Sci. 77: 4769-4773.

(29) Gerisch, G. 1968. Zell Aggregation bei Dyctiostelium. Curr. Top. Dev. Bio. 3: 157-197.

(30) Goldman, R.D., and Knipe, D. 1972. Functions of cytoplasmic fibers in non-muscle cells. Cold Spring Harbor Symp. Quant. Biol. 37: 523-534.

(31) Heaysman, J., and Pegrum, S. 1973. Early contacts between fibroblasts. An ultrastructural study. Exp. Cell Res. 78: 71-78.

(32) Heuser, J.E., and Kirschner, M. 1980. Filament organization revealed in platinum replicas of freeze-dried cytoskeletons. J. Cell Biol. 86: 212-234.

(33) Hoffman, P.N., and Lasek, R.J. 1975. The slow component of axonal transport. J. Cell Biol. 66: 351-366.

(34) Hynes, R.O. 1981. Relationships between fibronectin and the cytoskeleton. In Cell Surface Reviews, eds. G. Poste and G. Nicolson, vol. 7. New York: Alan R. Liss, in press.

(35) Hynes, R.O., and Destree, A.T. 1978. Relationship between fibronectin and actin. Cell 15: 875-886.

(36) Hynes, R.O., and Destree, A.T. 1978. 10 nm filaments in normal and transformed cells. Cell 13: 151-163.

(37) Ishikawa, H.; Bischoff, R.; and Holtzer, H. 1968. Mitosis and intermediate sized filaments in developing skeletal muscle. J. Cell Biol. 38: 538-555.

(38) Jorgenson, A.O.; Subrahmanyan, L.; Turnbull, C.; and Kalnins, V.I. 1976. Localization of the neurofilament protein in neuroblastoma cells by immunofluorescent staining. Proc. Nat. Acad. Sci. 73: 3192-3196.

(39) Lasek, R. 1981. Axonal transport. In The Cytoskeleton and the Architecture of Nervous Systems, eds. R. Lasek and M. Shelanski. Cambridge, MA: Neurosciences Research Program, in press.

(40) Lasek, R.J.; Solomon, F.; and Brinkley, B.R. 1981. Organizing centers: the form and transport of cell skeletons. In the Cytoskeleton and the Architecture of Nervous Systems, eds. R. Lasek and M. Shelanski. Cambridge, MA: Neurosciences Research Program, in press.

(41) Lazarides, E. 1980. Intermediate filaments as mechanical integrators of cellular space. Nature 283: 249-255.

(42) Lenk, R., and Penman, S. 1979. The cytoskeletal framework and poliovirus metabolism. Cell 16: 289-301.

(43) Loor, F. 1976. Cell surface design. Nature 264: 272-273.

(44) Luby, K., and Porter, K.R. 1980. The control of pigment migration in isolated erythrophores of Holocentrus ascensionis (Osbeck). I. Energy requirements. Cell 21: 13-23.

(45) Lux, S.E. 1979. Spectrin-actin membrane skeleton of normal and abnormal red blood cells. Semin. Hematol. 16: 21-51.

(46) MacPerson, I., and Montagnier, L. 1964. Anchorage-dependent growth of normal cells. Virology 23: 291-294.

(47) McClain, D.A., and Edelman, G.M. 1980. Density-dependent stimulation and inhibition of cell growth by agents that disrupt microtubules. Proc. Nat. Acad. Sci. 77: 2748-2752.

(48) Nicholls, J.G., ed. 1979. The Role of Intercellular Signals: Navigation, Encounter, Outcome. Weinheim, New York: Verlag Chemie.

(49) Nicolson, G.L. 1976. Transmembrane control of the receptors on normal and cancer cells. I. Cytoplasmic influence over cell surface components. Biochim. Biophys. Acta 457: 57-108.

(50) Nicolson, G.L. 1976. Transmembrane control of the receptors on normal and cancer cells. II. Surface changes associated with transformation and malignancy. Biochim. Biophys. Acta 458: 1-72.

(51) Oliver, J.M.; Ukena, T.E.; and Berlin, R.D. 1974. Effects of phagocytosis and colchicine on the distribution of lectin-binding sites in cell surfaces. Proc. Nat. Acad. Sci. 71: 394-398.

(52) Otto, A.M.; Zumbe, A., Gibson, L.; Kubler, A.-M.; and Jimenez de Asuce, L. 1979. Colchicine enhances the effect of growth factors on 3T3 cells. Proc. Nat. Acad. Sci. 76: 6435-6438.

Cell Interactions and the Cytoskeleton   145

(53) Pollard, T.D., and Fujiwara, K. 1976. Fluorescent antibody localization of myosin in the cytoplasm, cleavage
furrow, and mitotic spindle of human cells. J. Cell
Biol. 71: 848-875.

(54) Sanes, J.S.; Marshall, L.M.; and McMahon, U.J. 1978.
Reinnervation of muscle fiber basal lamina after removal
of myofibres. J. Cell Biol. 78: 176-198.

(55) Sherline, P., and Shiavone, K. 1977. Immunofluorescent
localization of high molecular weight proteins along
intracellular microtubules. Science 198: 1038-1040.

(56) Singer, I.I. 1979. The fibronexus: a transmembrane
association of fibronectin-containing fibers and bundles
of 5 nm microfilaments in hamster and human fibroblasts.
Cell 16: 675-685.

(57) Solomon, F. 1980. Neuroblastoma cells recapitulate
their detailed neurite morphologies after reversible
microtubule disassembly. Cell 21: 333-338.

(58) Solomon, F.; Magendantz, M.; and Salzman, A. 1979.
Identification with cellular microtubules of one of the
co-assembling microtubule-associated proteins. Cell 18:
431-438.

(59) Weiss, P. 1958. Cell contact. Int. Rev. Cytol. 7:
1217-1221.

(60) Wolosewick, J., and Porter, K.R. 1979. Microtrabecular
lattice of the cytoplasmic ground substance. Artifact
or reality. J. Cell Biol. 82: 114-139.

(61) Zieve, G., and Solomon, F. 1980. Identification of a
120K dalton protein associated with microtubules of the
mitotic spindle. J. Cell Biol., in press.

Neuronal-glial Cell Interrelationships, ed. T.A. Sears, pp. 147-158.
Dahlem Konferenzen 1982. Berlin, Heidelberg, New York: Springer-Verlag.

Signalling Between Neuronal and Glial Cells

R. K. Orkand
Dept. of Physiology and Pharmacology, University of Pennsylvania
School of Dental Medicine, Philadelphia, PA 19104, USA

Abstract. The maintenance of normal nervous system function
appears to involve continuous communication between neurons
and their associated cells, the neuroglia. Many possible
mechanisms exist for the transfer of information between the
two cell types which are separated only by narrow, 200Å wide,
extracellular clefts. There is increasing evidence that
several of these are used. These include: 1) Activity in
one cell varies the ionic composition of the extracellular
cleft and this change is detected by the other cell. The release of K^+ from neurons to depolarize glia is one such
example. 2) Neurotransmitter agents, including amino acids
and their derivatives, biogenic amines and peptides, used in
neuronal signalling also act at membrane or intracellular
loci of the nearby glial cells and alter their metabolism.
3) Macromolecules, especially proteins, are exchanged between
the cells by phagocytosis or a combination of exocytosis and
endocytosis. The accumulation in the extracellular clefts of
some substances released from neurons serves as a signal to
the glial cells to clear that substance from the extracellular space and maintain the constancy of the extracellular
fluid. Although there is evidence for other functional interactions between the cells, for the most part the specific
signalling processes are poorly defined and their physiological significance obscure.

INTRODUCTION

The idea that glial cells and neurons are functionally related

had its origins in nineteenth century light microscopic stu-

dies; it has received great impetus from modern electron mi-

croscopy (6,11,23). Close anatomical relations have suggested

close physiological relations. Neuron-glia interactions

during ontogeny, injury, and repair are properly discussed in

other papers of this volume. I shall consider processes
which may serve as signals to coordinate activity of neurons
and glia that are aimed at the maintenance of normal nervous
system function.

Functional coordination of activity in neurons and glial
cells requires the existence of cellular mechanisms whereby
one cell type can detect and respond to a change in activity
in the other. What evidence is there that neuronal and glial
activity is coordinated in the normal nervous system? What
mechanisms exist for the transfer of a signal from one cell
to the other? Experimental evidence is accumulating for processes which might serve to link activity in neurons and glia.
However, specific hypotheses for their functional interdependence remain largely speculative.

NEURON-GLIAL SIGNALS
Neurons and glia, although closely apposed, remain anatomically
distinct. Some glial cells can communicate with one another
through gap junctions which permit the exchange of ions and
small molecules (16). Neurons communicate primarily at specialized chemical synaptic contacts and occasionally via electrical connections at gap junctions. In Mugnaini's paper (this
volume), the membrane specializations between glial cells and
neurons are detailed and discussed. Over most of their adjacent surfaces neurons and glia are separated by the narrow
intracellular clefts which constitute the functional microenvironment of the nervous system. The present discussion is
focused on the possibility that glial cells and neurons communicate by altering the ionic composition of their mutual
bathing solution by releasing biologically active small molecules and by transferring macromolecules.

Potassium
The membrane potential of glial cells arises from the unequal
distribution of K^+ across the membrane. That is, in the range
of $[K^+]_o$ found in the nervous system, the glial membrane behaves as a K^+ electrode according to the Nernst relation
$Vm = RT/F \ln [K^+]_o / [K^+]_i$ (12). By comparison, the membrane

potential of neurons is not very sensitive to changes in
$[K^+]_o$ at normal levels (16). Nerve impulses or excitatory
synaptic potentials result from a small influx of Na^+ or
Ca^{2+} into the neuron and a similar efflux of K^+ into the
narrow system of clefts which constitute the functional
extracellular space of the nervous system (11). These clefts
have an ionic composition comparable to that in the cerebrospinal fluid: high Na^+ and low K^+ and Ca^{2+} (4). The ionic
shifts during neuronal activity, although amounting to only
a few picomoles of ions per cm^2 of nerve membrane, are
sufficient to produce significant increases in $[K^+]_o$ (17) and
decreases in $[Ca^{2+}]_o$ (15) which last many seconds. The magnitude
of the ionic change is a function of the intensity and spatial
distribution of the neuronal activity as well as the homeo-
static processes tending to restore the ionic distribution.
With intense neuronal activity, $[K^+]_o$ can increase more than
fivefold (20) and $[Ca^{2+}]_o$ might decrease by one-third (15).
The increase in $[K^+]_o$ produces a marked depolarization of the
glial cells. We do not know if the glial cell is sensitive to
the changes in $[Ca^{2+}]_o$. The depolarization of the glial membrane
spreads passively through the glial syncytium by way of the
gap junctions. At any one time, the glial membrane potential
is a good indication of the amount of neuronal activity in the
vicinity. The release of K^+ from active neurons and its accumulation in the extracellular clefts could serve as a signal
to coordinate neuronal and glial activity (10,19).

Increases in $[K^+]_o$ can affect glial metabolism (7). In
one series of experiments, fluorometric techniques were used
to measure levels of NADH in intact amphibian optic nerves
containing unmyelinated axons and glial cells and also in
optic nerves consisting solely of glial cells. The latter
were obtained by enucleating animals and waiting two months
for all the axons to degenerate. In both these preparations,
an increase in $[K^+]_o$ produced a decrease in the level of NADH
(16). In other experiments, using the same preparations, an
increase in glial uptake of glucose and an increased

incorporation of labeled glucose into organic and amino acids
resulted from an increase in $[K^+]_o$ (18). The levels of $[K^+]_o$
necessary to produce these effects were similar to those
found to result from nerve activity (20). The release of K^+
from neurons can serve to adjust glial metabolism to the overall level of neuronal activity. It could be a signal for glial
cells to release metabolic substrates required by neurons or to
remove metabolic waste products released from the neurons.

Potassium released from active neurons might also serve to
signal glial cells to contribute to ion homeostasis in the
extracellular clefts. Although the present discussion centers
on control of K^+, one should appreciate that the function of
neurons depends not only on K^+ but also on Na^+, Ca^{2+}, Cl^-, and
H^+ in the microenvironment and that their movements are all
linked to that of K^+ (15). One hypothesis proposes that a
rise in $[K^+]_o$ serves as a signal for the glial cells to restore the K^+ gradients. At least two processes contribute to
this restoration: active transport and spatial buffering (5).

Active Transport

There is no doubt that active transport plays a role in the
dispersal of K^+ following its accumulation during neuronal
activity (15,21). The question is, what roles do glial cells
play in this dispersal? Electrogenic sodium pumps in neurons
are stimulated by the rise in both $[Na^+]_i$ and $[K^+]_o$. Glial cell
membranes have an electrogenic strophanthidin-sensitive Na^+ pump
which requires K^+ (21). The primary stimulus for this pump is
an increase in $[Na^+]_i$. An increase in $[K^+]_o$ does not appear, by
itself, to stimulate active electrogenic transport in glial cells.
If it did, one would not observe the usual K^+ electrode behavior of
the glial membrane. It is possible that an increase in $[K^+]_o$
stimulates active transport of K^+ and an anion, either Cl^- or
HCO_3^-, which has no electrical manifestation. Such a process
would lead to some swelling of the glial cells. This has
not yet been demonstrated. K^+Cl^- inward movement might also
occur passively in the course of establishing a new Donnan

equilibrium. However, so far there is no indication that
glial cells have a significant permeability to Cl^-.

Spatial Buffering

Local K^+ accumulation produces a depolarization of glial
cells which decays electrotonically with distance throughout
the glial syncytium. In the Necturus optic nerve the
electrical space constant (the distance for the potential to
fall to about 37%) is about 0.8 mm (Tang, unpublished).
(Comparable data is not available for other glia.) The potential difference produces current loops through the glial
cytoplasm and extracellular clefts. Because of the high se-
lectivity of the glial membrane for K^+, this current is
carried by K^+ into the glia at the point where K^+ accumulates
and by K^+ out of the glia some distance away. The return
current in the extracellular cleft is carried mostly by Na^+
and Cl^-, the predominant ions of the extracellular fluid.
Such currents tend to even out the concentrations in the extracellular clefts and may play a substantial role in the
dispersal of K^+ over distances greater than a few hundred
microns (15,17). The K^+ released from the neurons appears to
serve as a signal to the glial cells to equalize the K^+ in
the extracellular space.

Small Molecules

Neurons release a variety of neurotransmitters, neuromodula-
tors, and hormones into the extracellular space. These substances provide specific signals in neuron-to-neuron communication. Glial cells, by nature intimately apposed to neurons,
are also interwoven into synaptic regions and thus are exposed to these substances by diffusion from the synaptic
cleft. These neurosecretions could serve a dual role in neuronal and neuron-glial communication.

In glial cultures, the addition of catecholamines increases
levels of cAMP and promotes glycogenolysis (23). This observation is of special interest in view of the long-standing

hypothesis that glial cells serve as a source of glucose for
neurons (26). Other compounds, like acetylcholine and GABA,
have been found sometimes to depolarize glial cells (9).
However, in intact nervous systems it is difficult to distinguish between a direct effect of these neurotransmitters on
glial cells and an effect on neurons with subsequent release
of K^+ producing the glial depolarization. In our laboratory
(Tang & Orkand unpublished), we have found that glutamate depolarizes glial cells at relatively low concentrations
($10^{-5}M$) in the all-glial optic nerve.

In a series of papers, Villegas (24) has reported a complex
cholinergic interaction between squid axons and their surrounding Schwann cells. Nerve impulses are reported to lead
to a hyperpolarization of the Schwann cell which is mimicked
by carbachol, prolonged by eserine, and blocked by curare and
α-bungarotoxin, suggesting that it results from a release of
acetylcholine. Because the acetylcholine content is higher
in the Schwann cell than the axoplasm and the axolemma has
high acetylcholinesterase activity, Villegas suggests that
the acetylcholine is released from the Schwann cell itself.
In this preparation, the effect of l-glutamic acid is first
to hyperpolarize and then to depolarize the Schwann cell.
Since the hyperpolarization is blocked by α-bungarotoxin, it
is suggested that l-glutamic acid has two effects: first to
release acetylcholine from the Schwann cell, and second, to
directly depolarize the Schwann cell membrane. The critical
measurements of changes in glial membrane potential in these
studies are accomplished through the difficult technique of
making repeated microelectrode penetrations of a very thin
(0.2-0.6 micron) Schwann cell sheath. Because of the great
importance of these findings in establishing a whole new type
of neuron-glial interaction, it is important to have them confirmed with less demanding techniques.

Only a small fraction of the many biologically active small
molecules found in the central nervous system have been

examined for their ability to produce biochemical or electrophysiological effects on glial cells. More than two dozen
amino acid derivatives, monamine compounds and peptides are
being seriously considered as neurotransmitters, neuromodula-
tors, or hormones. Many of these compounds may play a role in
neuron-glia or glia-neuron communication in addition to their
proposed role in neuron-neuron signalling.

Glial cells are capable of active uptake and accumulation of
neurotransmitters like GABA, glutamate, and the biogenic amines
(7,16). Glial fluxes of GABA have been studied in some detail
in mammalian sympathetic ganglia (2,14). The studies suggest
that glial cells do not signal back to neurons via GABA but instead fulfill a simple clearance function for extracellular GABA.
The same system is present in mammalian brain but is secondary
to the more rapid neuronal uptake system in terms of inactivating
exogenous GABA (7). No evidence has yet been adduced to show
that either glial or neuronal transport systems affect the duration of GABA-mediated synaptic inhibitory potentials; their role
is presumably to maintain steady-state interstitial levels below
those capable of activating the GABA receptors (8). Once inside the glial cells it is likely that the neurotransmitters affect
metabolic reactions. Thus, the release of these neurotransmitters not only transmits a neuron-neuron signal but also
initiates a metabolic change in the glial cells. In addition,
the rise in concentration of the transmitter in the extracellular clefts serves as a signal to the glial transport
system to lower its extracellular concentration.

Macromolecules
Experiments with the squid giant axon and its associated
Schwann cell sheath suggest that proteins made in the Schwann
cell are transferred to the axoplasm (13,23). The addition
of radiolabeled amino acids leads to the appearance of labeled protein in both the sheath and the axoplasm. The protein does not appear to be made by the axon because extruded
axoplasm contains no ribosomal RNA and cannot synthesize protein from amino acids. Moreover, internal perfusion of the

intact axon with amino acids does not lead to protein synthesis. The amino acids must be added to the sheath for labeled
protein to appear in the axoplasm. The transfer of protein
is not affected by nerve impulses but is inhibited by reduced
calcium or added magnesium. Possible mechanisms for the proposed transfer include a combination of Schwann cell exocytosis and axonal endocytosis and the phagocytosis of small
glial extrusions which invaginate the axolemma. The ability
of glia and axons to exchange macromolecules greatly increases
the range of messages which might pass between the two cell
types.

About 30 years ago, Levi-Montalcini and Hamburger discovered
a protein which promoted neurite outgrowth from neurons in
tissue culture. The substance, termed nerve growth factor,
is found in a variety of tissues, especially salivary glands.
It is possible that glial cells can produce nerve growth
factor; they do produce other nerve growth factor-like agents
(22). In systems where nerve growth factors are required for
the survival of neurons, glial cells may release these essential substances.

Results in Search of a Signal
In our own experiments with enucleated amphibians (16), the
glial cells in the optic nerve remain morphologically intact
for about two months after all the axons have degenerated.
In succeeding months the glial cells accumulate masses of
dense bodies and lipid droplets and decrease in volume. The
cells remain for at least a year, but in animals kept two
years we have been unable to find the 'glial' nerve. The
presence of neurons appears necessary for the maintainance
of normal glial cells.

In a study of acetylcholine release from Schwann cells at denervated neuromuscular junctions, Bevan, Miledi, and Grampp
(1) found that actinomycin D which blocks DNA dependent RNA
synthesis suppresses the appearance of Schwann cell release

of acetylcholine which normally follows nerve terminal degeneration. The results are compatible with the idea that
Schwann cells are capable of synthesis and release of acetylcholine, but this activity is either repressed by a substance released from normal neurons or is initiated by an
activating agent released from degenerating terminals.
Either explanation involves the transfer of a signal from
the nerve terminal to the surrounding Schwann cell.

The role of axons in providing a signal for mitogenesis and
myelin production by Schwann cells will be discussed by
others in this workshop.

Acknowledgement. Supported in part by USPHS Grant NS-12253.

REFERENCES

(1)   Bevan, S.; Miledi, R.; and Grampp, W. 1973. Induced
      transmitter release from Schwann cells and its
      suppression by actinomycin D. Nature New Biol. 241:
      85-86.

(2)   Bowery, N.G.; Brown, D.A.; and Marsh, S. 1979. γ-Amino-
      butyric acid efflux from sympathetic glial cells: effect
      of "depolarizing" agents. J. Physiol. 293: 75-101.

(3)   Brown, D.A.; Collins, G.G.S.; and Glavan, M. 1980. In-
      fluence of cellular transport on the interaction of amino
      acids with γ-aminobutyric acid-receptors in the isolated
      olfactory cortex of the guinea-pig. Br. J. Pharmac. 8:
      251-262.

(4)   Cohen, M.W.; Gerschenfeld, H.M.; and Kuffler, S.W.
      1968. Ionic environment of neurons and glial cells
      in the brain of an amphibian. J. Physiol 197: 363-380.

(5)   Coles, J.A., and Tsacopoulos, M. 1979. Potassium ac-
      tivity in photoreceptors, glial cells and extracellular
      space in the drone retina: changes during photostimula-
      tion. J. Physiol. 290: 525-549.

(6)   Glees, P. 1955. Neuroglia Morphology and Function.
      Springfield, Il: C.C. Thomas.

(7)   Hertz, L. 1977. Drug-induced alterations of ion distri-
      bution at the cellular level of the central nervous
      system. Pharmacol. Rev. 29: 35-65.

156   R.K. Orkand

(8) Horwitz, I.S., and Orkand, R.K. 1980. GABA inactivation
     at the crayfish neuromuscular junction. J. Neurobiol. 11:
     447-458.

(9) Krnjević, K., and Schwartz, S. 1967. Some properties of
     unresponsive cells in the cerebral cortex. Exp. Brain
     Res. 3: 306-319.

(10) Kuffler, S.W. 1967. Neuroglial cells: physiological
     properties and a potassium mediated effect of neuronal
     activity on the glial membrane potential. Proc. R. Soc.
     B. 168: 1-21.

(11) Kuffler, S.W., and Nicholls, J.G. 1966. The physiology
     of neuroglial cells. Ergebn. Physiol. 57: 1-90.

(12) Kuffler, S.W.; Nicholls, J.G.; and Orkand, R.K. 1966.
     Physiological properties of glial cells in the central
     nervous system of amphibia. J. Neurophysiol. 29: 768-
     787.

(13) Lasek, R.J.; Gainer, H.; and Barker, J.L. 1977. Cell-
     to-cell transfer of glial proteins to the squid giant
     axon. The glia-neuron protein transfer hypothesis.
     J. Cell Biol. 74: 501-523.

(14) Minchin, M.C.W., and Iversen, L.L. 1974. Release of
     {3H}gamma-aminobutyric acid from glial cell in rat dorsal
     root ganglia. J. Neurochem. 23: 533-540.

(15) Nicholson, C. 1980. Dynamics of the brain cell micro-
     environment. Neurosci. Res. Prog. Bull. 18: 177-322.

(16) Orkand, R.K. 1977. Glial cells. In Handbook of Physiol-
     ogy, section 1: The Nervous System: Cellular Biology of
     Neurons, eds. J.M. Brookhart, V.B. Mountcastle, F.R. Kandel,
     and S.R. Geiger, vol. 1, part 2, pp. 855-875. Bethesda, MD:
     American Physiological Society.

(17) Orkand, R.K.; Nicholls, J.G.; and Kuffler, S.W. 1966.
     Effect of nerve impulses on the membrane potential of
     glial cells in the central nervous system of amphibia.
     J. Neurophysiol. 29: 788-806.

(18) Salem, R.D.; Hammerschlag, R.; Bracho, H.; and Orkand,
     R.K. 1975. Influence of potassium ions on accumulation
     and metabolism of {14C}glucose by glial cells. Brain
     Res. 86: 499-503.

(19) Somjen, G.G. 1979. Extracellular potassium in the
     mammalian central nervous system. Annu. Rev. Physiol.
     41: 159-177.

(20) Syková, E., and Orkand, R.K. 1980. Extracellular
     potassium accumulation and transmission in frog spinal
     cord. Neurosci. 5: 1421-1428.

(21)	Tang, C.-M.; Cohen, M.W.; and Orkand, R.K. 1980.
	Electrogenic pumps in axons and neuroglia and extra-
	cellular potassium homeostasis. Brain Res. 194: 283-286.

(22)	Varon, S., and Bunge, R. 1978. Trophic mechanisms in
	the peripheral nervous system. Annu. Rev. Neurosci. 1:
	327-361.

(23)	Varon, S.S., and Somjen, G.G. 1979. Neuron-glia inter-
	actions. Neurosci. Res. Progr. Bull. 17(1).

(24)	Villegas, J. 1978. Cholinergic properties of satellite
	cells in the peripheral nervous system. In Dynamic
	Properties of Glia Cells, eds. E. Schoffeniels et al.
	New York: Pergamon Press.

(25)	Watson, W.E. 1974. Physiology of neuroglia. Physiol.
	Rev. 54: 245-271.

(26)	Wigglesworth, V.B. 1960. The nutrition of the central
	nervous system of the cockroach Periplaneta american L.
	The role of perineurium and glial cells in the mobiliza-
	tion of reserves. J. Exp. Biol. 37: 500-512.

Neuronal-glial Cell Interrelationships, ed. T.A. Sears, pp. 159-167.
Dahlem Konferenzen 1982. Berlin, Heidelberg, New York: Springer-Verlag.

Aspects of Receptor Systems Possibly Relevant to Neuron-Glial Interactions and Multiple Sclerosis

D. M. Fambrough
Dept. of Embryology, Carnegie Institute of Washington
Baltimore, MD 21210, USA

Abstract. Receptor mediated phenomena can be subdivided into
categories based upon the chemical nature of ligand or receptor, the cellular location of functional receptor sites, and
the mechanisms of transduction of ligand-receptor interactions
into cellular responses. These classifications reflect the
very broad range of receptor systems which may share only one
quality: specificity. A principle value of the classifications
is their utility as mental frameworks with which to approach
the analysis of specific interactions in biological systems.
At present, neuron-glial cell interactions are understood so
fragmentarily that a major effort is needed to determine which
elements of the interactions show specificity.

RECEPTORS AS MEDIATORS OF SPECIFIC INTERACTIONS

In its most general sense, the concept of receptor implies recognition or specificity. With few exceptions this specificity
involves a set of noncovalent interactions between receptor and
ligand molecules. As in the case of enzyme-substrate interactions,
the specificity usually results from molecular congruence: receptor and ligand can assume molecular conformations such that
multiple weak bonds form when the molecules collide with favorable orientation. Almost always this means that at least one
of the pair is a protein. (We know this empirically and also
could deduce it from general considerations of the structures
of biochemicals and the principles of inheritance and evolution.)
Frequently, ligand-receptor systems are involved in communication

and, by convention, the more mobile ligand corresponds to a
signal and the less mobile is considered the receptor. Especially in the contexts of cell-cell association and virus-cell
interactions we should avoid trivial semantic problems that
might result from insisting on ligand-receptor terminology.
After all, some "viral receptors" on vertebrate cells may be
the complex carbohydrate components of gangliosides (rather
mobile lipids in the plasma membrane), while the ligand is a
rigidly fixed coat protein of the virus particle, tailored
through evolution to display high-affinity binding sites for
the target cell gangliosides.

There are several ways in which we might classify receptor-ligand
systems: on the basis of the chemical nature of ligand or receptor, on the basis of the cellular location of the sites of
ligand-receptor interaction, and on the basis of the type of
mechanism by which the ligand-receptor interaction leads to the
appropriate consequences. It is worthwhile to examine ligand-
receptor systems in the context of each of these classification
schemes. Indeed, each of these schemes is really the construct
of experimentalists. These schemes are lists of possible answers to the experimentalists' questions: a) what is the chemical
nature of the ligand and of the receptor? b) where in (or on)
the cell does the ligand-receptor interaction occur? and c) how
does the interaction lead to the response? The answers to such
questions must be obtained before we have a satisfactory understanding of neuron-glial interactions or the molecular basis of
multiple sclerosis. In fact, our understanding of these matters
is so rudimentary that we are still faced with the job of determining in what ways ligand-receptor systems are involved. As
we explore various aspects of neuron-glial interaction, we should
bear in mind that specificity is the hallmark of ligand-receptor
systems.

CHEMICAL NATURE OF LIGANDS AND RECEPTORS
Hypothetically, almost any sort of ion or molecule could serve
as ligand in a ligand-receptor system in which the receptor was

proteinaceous. That this is so is supported by the example of
the immune system. Immunoglobulin molecules (which we can consider receptors for their target antigens) have been elicited
in experimental animals by appropriate immunization with a huge
variety of different haptens. The immunoglobulins share a common overall shape, but different immunoglobulins have different
amino acid sequences in their variable regions. These differences
in primary structure lead to differences in the molecular configuration of the antigen binding sites and thus to different
specificities. While most receptor systems probably share with
the immunoglobulins the characteristic that one member of the
ligand-receptor pair is a protein, various receptor systems can
be classified on the basis of the nature of the other member of
the pair. Identifying the chemical nature of this second member is a major step in the characterization of any receptor system. Its chemical nature will have a great deal to do with the
further analysis of the receptor system. It will affect the
choice of hypotheses as to how the system might work and it
will even affect many investigators' decisions as to whether
they are equipped to study the system.

In the systems with soluble ligands, the route to establishing
the chemical identity of ligand is usually straightforward:
develop a bioassay, use the assay to locate the ligand during
its extraction and purification, and then analyze the purified
ligand. (Depending on the nature of the assay and the ligand
this may take hours or years.) A much more difficult task is
the characterization of components in receptor systems involving
cell-bound or insoluble components. Such components are the
sort one might expect to mediate such phenomena as specific cell
associations. Efforts in this area have generally been uncon-
vincing. To this day it remains unclear what role cell surface
carbohydrates play in cell-cell interactions and whether soluble
"linkers" analogous to lectins are important. Whatever the
mechanisms of cell association, we imagine they must override
the repulsive force of negatively charged surfaces, and we know
that calcium and possibly other multivalent cations are essential

but quite non-specific facilitators. Investigators in this
area have been stymied by the problem of defining operationally
where and when specificity occurs during establishment of cell-
cell associations.

The characterization and isolation of receptors have been major
goals in pharmacology and biochemistry. A classical pharmacological approach has been to prepare chemical analogues of the
natural ligand and to study their efficacy, thus deriving inferences as to the nature of the ligand binding site. Bi-
functional synthetic ligands have been used to explore sites
in proximity to the receptor site. They have also been used to
label receptor molecules covalently so that the receptor may be
removed from its natural environment and its purification followed whether or not such a receptor still retains its ligand
binding site or its capacity to carry out a measurable ligand-
induced function. The isolation of receptors and further chemical characterization have been important in validating the concept that many receptors are proteins with receptor sites comparable to substrate binding sites of enzymes. Knowing the
chemical nature of the receptor may help investigators determine the mechanism of receptor function.

Another goal related to receptor isolation and characterization
is simplification of the ligand-receptor system with maintenance
of function. For example, studies on receptor systems in which
the response involves biosynthesis of cyclic nucleotide have
progressed largely through partial purification of responsive
plasma membrane. Reconstitution of functional ligand-receptor
systems from well-characterized molecular constituents is both
part of a proof of a particular mechanism of receptor function
and also a starting point for studies on some molecular details
of receptor function.

RECEPTOR LOCATION
Studies involving the use of radioisotopically labeled ligands
together with techniques of subcellular fractionation or

microscopy have established that receptors for some lipophilic ligands such as steroid hormones and thyroxin may reside in intracellular compartments. Receptors for more polar ligands (such as the proven neurotransmitters and most polypeptide hormones) are membrane-associated and the ligand-receptor interaction takes place at the cell surface, at least initially. Some of the cell-surface receptors mediate uptake of ligand (for example, the low density lipoprotein receptor, the transferrin receptor, and the asialoglycoprotein receptor (4,7,10)) and the ligand-receptor interaction may continue during the formation of endocytotic vesicles, resulting in temporary or permanent removal of the receptor from the plasma membrane. It is probably true that receptors which mediate rather massive uptake (such as the receptors mentioned immediately above) are reutilized by being recycled back to the plasma membrane while receptors for some of the polypeptide hormones (such as the epidermal growth factor receptor and the insulin receptor) may be destroyed after internalization, leading to "down regulation" of the responding cells. There is speculation that some ligand-receptor complexes (for example, nerve growth factor and the NGF receptor) may be conveyed to another site, such as the nuclear membrane, where the complex acts as an effector of subsequent events (11).

There is abundant evidence that cell surfaces are organized so that different functions are localized in appropriate places. This is most precisely determined for the nicotinic cholinergic receptors of skeletal muscle fibers (3), which are highly concentrated in the juxtaneural portions of the post-synaptic membrane at neuromuscular junctions. Other examples include the molecules that form various epithelial cell junctions and that perform transport functions of various polarized cells. Some receptors related to uptake mechanisms (such as the LDL receptor) are thought to occur preferentially in coated pits, while other receptors (such as the semliki forest virus receptor (6) and the immunoglobulin receptors of the fetal gut (8)) are distributed over the cell surface until interaction with ligand

triggers the movement of ligand-receptor complexes into coated
pits prior to uptake.

Analysis of the developmental regulation of receptors has contributed some important insights. Over 75 years ago Harrison
(5) attempted to answer the question: is functional validation
an important aspect of the establishment of correct cell-cell
interactions? Anaesthetized frog embryos were grown to the
swimming tadpole stage and then returned to normal conditions,
whereupon they quickly began to swim and show normal reflexive
responses. A refined, specific case of recent vintage was the
demonstration that the clustering of cholinergic receptors at
neuromuscular junctions and the elaboration of synaptic specializations are independent of cholinergic transmission (1). Recent observations by McMahan and co-workers on mechanisms of
reinnervation of skeletal muscle have contributed a new dimension (2,9). They have demonstrated that accurate reinnervation
and re-elaboration of synaptic structure may be mediated through
rather permanently fixed components of the extracellular matrix
at the original site of synapsis, so that both pre- and postsynaptic structures can be regenerated by their respective cells
in the absence of the other member of the pair and without the
participation of glial elements. This "instructive" role of
extracellular matrix suggests the possibility that the formation
of synaptic connections involves placement in the synaptic
cleft of "ligands" which then might continuously stimulate pre-
and postsynaptic receptors which mediate differentiation and
maintenance of pre- and postsynaptic organizations.

MECHANISMS OF RECEPTOR FUNCTION
Specificity may be the only property common to all of the
receptor-mediated phenomena considered in this workshop. However, it is worthwhile to consider that in most receptor systems the binding of ligand to receptor triggers other events.
Receptor systems can be classified on the basis of the type
of molecular events that follow ligand-receptor interaction.

For this discussion we might begin with a classification such
as the following.
1. Ligand-receptor complexes without transducer function;
    a. cell-cell adhesion
            possibly including tight junctions,
            desmosomes
    b. cell-substrate adhesion
            interactions of cells with their extracellular
            matrix
    c. viral attachment
2. Ligand-receptor complexes which carry out a function at
    the cell surface
    a. ligand-receptor complexes leading to ion channel
        formation by conformation change
    b. ligand-receptor complexes mediating active uptake or
        extrusion of another substance
    c. ligand-receptor complexes causing generation of a
        second messenger by an enzymatic mechanism
3. Ligand binding triggering interiorization of receptor
    a. receptor-mediated endocytosis of ligand as an uptake
        mechanism
    b. ligand-receptor complex (or some fragment of one or
        the other, or both) serving as an internal ligand in
        another step in the response
    c. interiorization terminating a process mediated by
        the free receptor.

Clearly, when the receptor plays a role as <u>transducer</u>, the subsequent response can be a complex function of <u>receptor</u> <u>occupancy</u>.
Even the concept of receptor occupancy becomes blurred when we
consider interiorization of ligand-receptor complexes and so on.
Furthermore, the time-courses of responses may not correlate
with the time-course of receptor occupancy.

MODULATION OF RECEPTOR FUNCTION
The number of receptor sites of a cell will depend upon biosynthetic production of receptors by the cell and upon the rate of

degradation of receptors. Perturbations that change either of
these steps in receptor metabolism will result in a change in
the total receptor number. In the case of acetylcholine receptors of skeletal muscle all four of the major possible directions
of perturbation can occur (3). Denervation leads to accelerated
biosynthesis and accumulation of excess receptors. Electrical
stimulation of denervated muscle leads to suppression of receptor production and a loss of receptors through continued degradation. Receptor degradation is accelerated by anti-receptor
antibodies without compensatory acceleration of production, leading to a reduced number of receptor sites. At about the time
of innervation during development (or shortly thereafter), the
rate of receptor degradation is greatly decreased, contributing
to the accumulation of receptors at the neuromuscular junction.
As mentioned earlier, the down-regulation of receptors by polypeptide hormones may represent other cases in which accelerated
internalization and degradation are not fully compensated by
increased production.

POSSIBLE ROLES OF RECEPTORS IN NEURON-GLIAL INTERACTION AND
IN MULTIPLE SCLEROSIS
Exploration of this topic was one goal of this workshop. The
following possibilities have occurred to me:
1. receptors mediating specific contact between glial and
   neuronal elements,
2. receptors mediating trophic support of neuronal elements by
   glia,
3. receptors on glial cells related to termination or modulation
   of nervous activity,
4. neuronal and glial cell surface molecules which may serve
   as viral receptors,
5. cell surface (host or viral) molecules related to disruption
   of normal glial-neuron interactions, and
6. cell surface molecules related to participation of immune
   systems in disease processes.
There may be additional categories to consider. It is clear
that further specification of the above categories is required.

Acknowledgements. This paper is dedicated to Mr. Taniguchi, founder of the Taniguchi Foundation, who has directed his philanthropy toward the goal of world peace through the strengthening of personal ties and facilitation of intellectual exchange.

REFERENCES

(1) Anderson, M.J., and Cohen, M.W. 1977. Nerve-induced and spontaneous redistribution of acetylcholine receptors on cultured muscle cells. J. Physiol. (London) 268: 757-773.

(2) Burden, S.J.; Sargent, P.B.; and McMahan, U.J. 1979. Acetylcholine receptors in regenerating muscle accumulate at original synaptic sites in the absence of the nerve. J. Cell Biol. 82: 412-435.

(3) Fambrough, D.M. 1979. Control of acetycholine receptors in skeletal muscle. Physiol. Rev. 59: 165-227.

(4) Goldstein, J.L.; Anderson, R.G.W.; and Brown, M.S. 1979. Coated pits, coated vesicles, and receptor mediated endocytosis. Nature 279: 679-685.

(5) Harrison, R.G. 1904. An experimental study of the relation of the nervous system to the development of musculature in the embryo of the frog. Am. J. Anat. 3: 197-200.

(6) Helenius, A.; Kartenbeck, J.; Simons, K.; and Fries, E. 1980. On the entry of Semliki Forest Virus into BHK-21 cells. J. Cell Biol. 84: 404-420.

(7) Octave, J.-N.; Schneider, Y.J.; Hoffman, P.; Tronet, A.; and Crichton, R.R. 1979. Transferrin protein and iron uptake by cultured rat fibroblasts. FEBS Lett. 108: 127-130.

(8) Rodewald, R. 1980. Distribution of IgG receptors in the small intestine of the young rat. J. Cell Biol. 85: 18-32.

(9) Sanes, J.R.; Marshall, L.M.; and McMahan, U.J. 1978. Reinnervation of muscle fiber basal lamina after removal of myofibers: differentiation of regenerating axons at original synaptic sites. J. Cell Biol. 78: 176-198.

(10) Steer, C., and Ashwell, G. 1980. Studies on a mammalian hepatic binding protein specific for asialoglycoproteins. J. Biol. Chem. 255: 3008-3013.

(11) Yanker, B.A., and Shooter, E.M. 1979. Nerve growth factor in the nucleus: interaction with receptors on the nuclear membrane. Proc. Natl. Acad. Sci. (U.S.A.) 76: 1269-1273.

Group on Maintenance

Standing, left to right: Jack Griffin, Alan Davison,
Herman Rohrer, Bernard Droz, Günther Fischer, David Brown,
Doug Fambrough, John Treherne.
Seated: Dick Orkand, Rhona Mirsky, John Nicholls, Pat Walicke,
Dick Bunge.

Neuronal-glial Cell Interrelationships, ed. T.A. Sears, pp. 169-202.
Dahlem Konferenzen 1982. Berlin, Heidelberg, New York: Springer-Verlag.

Maintenance
State of the Art Report

P. A. Walicke, Rapporteur
D. A. Brown, R. P. Bunge, A. N. Davison, B. Droz,
D. M. Fambrough, G. Fischer, J. W. Griffin, R. Mirsky,
J. G. Nicholls, R. K. Orkand, H. Rohrer, J. E. Treherne

INTRODUCTION

In order to understand the effects of disease on a tissue, it
is first necessary to understand its normal physiological
working. This discussion will center on our knowledge of the
importance of the normal interactions of neuronal and glial
cells, and on how the interrelationship between these cells
is maintained. It is organized around the maintenance of
various definable structures in nervous tissue, starting from
simple ion distributions progressing up to whole cells. One
of its chief values will probably be in mapping some of the
vast gaps in our knowledge of these normal interactions, and
in pointing out areas ready for fruitful investigation.

CYTOLOGICAL OVERVIEW

Before turning to the question of maintenance, it is first
necessary to delineate the cell types to be considered.
Especially if this discussion is to be of use in understanding pathological processes, careful distinction between

different cell types must be made, since different classes of
glia are preferentially affected by different diseases. Five
major types of glial cells will be considered: astrocytes,
oligodendrocytes, myelinating and nonmyelinating Schwann cells,
and enteric glial cells.

Although a wide variety of interactions between neuronal and
glial cells will be discussed, it is possible to classify
neuron-glial relationships under two general categories:
loose ensheathment and confining ensheathment or myelination
(19). Loose ensheathment is typified by the relationships
between astrocytes, non-myelinating Schwann cells, and enteric
glial cells and their respective neurons. The glial cells
surround and support the neuronal cells and axons, but leave a
surrounding extracellular space of about 200 Å through which
proteins and other constituents of the extracellular fluid can
pass relatively freely. Myelination represents a much more
specialized type of interaction between oligodendrocytes and
myelinating Schwann cells and their axons. A highly specialized membranous extension of the glial cell wraps itself
around the axon and forms a unique cellular junction with it
in the paranodal region (Mugnaini, this volume), which limits
access to the periaxonal space. The myelin formed by Schwann
cells and oligodendrocytes is not identical; for example, there
are differences in its periodicity and in the presence of
microvilli impinging on the nodal space. Perhaps most striking is that each Schwann cell wraps only one internode, while
a single oligodendrocyte can serve 35 or more.

The morphological features of these various types of glia have
been described in detail in numerous previous publications.
More recently, attention has focused on delineating biochemical and antigenic markers for the various cell types which can
be used to identify them during development, in pathological
states and in tissue culture where morphology is less depend-
able. Presumably such markers also help designate some of the
unique and important functions of the different cell types.

A brief list is presented in Table I. Even this abbreviated
list defines some subsets of the morphological glial classes,
such as the subclasses of astrocytes marked by the Cl and Ml
antigens (Schachner et al., this volume). The further use of
monoclonal antibodies against surface and intracellular antigens is likely to characterize further glial subclasses.

Table I
Biochemical and Antigenic Markers Distinguishing
Glial Cell Types

Cell Type	Marker
Astrocyte	Glial Fibrillary Acidic protein
	RAN-2 (rat only, also on ependymal and leptomenin-geal cells) (79)
subclasses	Cl, Ml (Schachner et al., this volume)
Oligodendrocyte	Galactocerebroside
	Sulfatide
	Myelin basic protein
	Myelin-associated glycoprotein (MAG)
	2',3' CNPase
	Glucose 6-phosphate dehydrogen-ase
subclasses	O_1, O_2, O_3, O_4 (Schachner et al., this volume)
Nonmyelinating Schwann cells	RAN-1 (rat only) (79)
Myelinating Schwann cells	RAN-1 (rat only) (79)
	Myelin glycoprotein (P_0)
	Myelin-associated glycoprotein (MAG)
	Myelin basic protein (P_1,P_2)
	Galactocerebroside
	Sulfatide
Enteric glial cells	RAN-1 (rat only)
	Glial fibrillary acidic pro-tein

Although classically most attention has focused on the interrelationships of neurons and glia, the interactions of glial
cells with each other are also of interest. Anatomically,

numerous gap junctions have been observed on CNS glial cells,
between individual astrocytes, and between astrocytes and
oligodendrocytes (Mugnaini, this volume). Physiologically,
there is evidence of electrical coupling between CNS glia in invertebrates (57) and amphibia (56) and between astrocytes in mammals (35). Dye coupling, with injection of lucifer CH, has been
observed between protoplasmic astrocytes in slices of guinea pig
neocortex (Gutnick, Connors, and Ranson, unpublished observation)
and in Necturus optic nerve (Tang and Orkand, unpublished observation). Physiological coupling or dye passage between astrocytes and oligodendrocytes has not yet been demonstrated. The
number of identified recordings from oligodendroglia, however,
are very few, and it is too early to make any conclusions regarding the physiological and functional properties of these connections. Thus, the glial cells of the CNS appear to form a large syncytium, possibly including both astroglia and oligodendroglia
cytoplasm within it. In contrast, myelinating Schwann cells
are not known to be in direct contact with each other in the
normal myelinated nerve. This difference in the degree of
coupling in central and peripheral elements could be important
in some functional and pathological contexts.

Questions:
1) Are any of the antigenic molecules defined so far important
in disease, particularly as target molecules?
2) Do known differences in antigens or cellular coupling explain the different susceptibility of Schwann cells and oligodendrocytes to the pathological process of multiple sclerosis?

BIOCHEMISTRY AND TURNOVER OF MYELIN
Considerable attention has been given to the biochemical characterization of CNS myelin. Compared to most cellular membrane preparations, myelin contains more lipid and less protein, less lecithin, and a higher proportion of galactolipid.
It is also somewhat noteworthy for its lower content of glycoproteins, which could mean fewer recognition sites and fewer

antigens on its membranes. It is not clear whether the remainder of the oligodendrocyte membrane differs significantly from
the composition of myelin, since no comparable preparation of
highly purified and characterized somal membrane exists.
Given the frequent observation of areas of different composition spatially maintained on the membranes of other cell types,
however, there is no reason not to assume that the remainder
of the oligodendrocyte may contain substantially different
proteins and glycoproteins from those characterized on myelin.

With the use of probes and radioactive ligands (15), detailed
models of the ultrastructure of the myelin unit-membrane of
the central nervous system have been constructed (71), though
some controversy still remains. High molecular weight proteins (e.g., Wolfgram protein) are suggested to be on the outer
myelin surface. Proteolipid protein probably transverses the
membrane and confers stability through covalently bound long
chain fatty acids. It appears to contain a pore through its
center, suggesting it may be involved in ion movements. In
these biochemical studies, myelin basic protein appears to be
on the inner part of the unit membrane, relatively inacces-
sible to extracellular factors, though immunohistochemical
studies give differing results. In the bimolecular lipid
phase, sulfatide may be concentrated on the side close to the
basic protein and is presumably attracted electrostatically.
Phospholipids are arranged with head groups interacting with
the protein layers and with their hydrophobic long chain fatty
acids penetrating into the lipid phase. There is asymmetrical
distribution of the lipids, with, for example, more cholesterol on the outer surface (25). In normal-appearing areas of
multiple sclerosis white matter, changes in lipid composition,
such as reductions in phospholipid polyunsaturated fatty acids,
have been suggested to demonstrate a myelin membrane disorder
(88). However, it seems quite likely that such changes in
lipid composition could be due to on-going pathological
change, invisible to the naked eye. Thus, increased hydrolase
activity and cholesteryl ester formation is associated with

the alterations in lipid composition. In other studies,
myelin isolated from normal-appearing white matter in multiple
sclerosis has a normal composition (87). It would seem that
any theories that propose a genetic or dietary alteration in
multiple sclerosis as a predisposing factor require more careful documentation and experimental verification (74).

In early studies of myelin turnover, the conclusion was
reached that myelin was an extremely stable substance. ^{14}C-
labelled cholesterol incorporated into the brain was observed
to persist as an intact molecule for years. This led to the
suggestion that myelin was metabolically inert, for cholesterol
is a prominent component of myelin. However, there were unexplained discrepancies. Isotopically labelled precursors
could be incorporated, in small proportions, into adult white
matter and myelin. Subsequent turnover was then slow. Exchange of labelled lipid molecules with cytoplasmic (presumably glial) organelles was demonstrated. Radioautographic
studies showed that molecules taken up on the outer and inner
parts of the sheath slowly penetrated into the deeper myelin lamallae (42). When turnover of individual components was examined,
proteolipid protein, cholesterol, and the galactolipids were
found to have a relatively high degree of stability ($t_{0.5} \sim$
100 days); the high molecular weight proteins turnover was
quicker ($t_{0.5} \sim$ 17 days); and a proportion of the myelin basic
protein was stable and the rest was metabolized more quickly
(8, 81).

Much of this work has been complicated by attempts to calcu-
late pool size, and to allow for growth and for reutilization
of complex as well as simple molecules. It appears that
lipids can be exchanged with glial cytoplasm and that while
phospholipids are catabolized, sterols and cerebrosides are
not. These lipids are reutilized in the slow remodelling of
the myelin sheath. The different rates of turnover of other
molecular species may be explained by their anatomical localization. For instance, inositol phospholipids, which metabolize

quite quickly, may be concentrated at or close to the node of
Ranvier or near the axoplasm. In tissue culture myelin sheaths
appear to change shape and there is some evidence to suggest
that myelin has a degree of fluidity (15). For example, con-
canavalin A receptors in the intra-period region are capable
cf diffusion in the plane of the membrane (67). Penetration
of labelled molecules and turnover of some myelin components
may be explained by the dynamic state of myelin macromolecular
complexes. In the case of disease or viral infection, myelin
turnover may be altered in complex ways with intracellular
metabolism of myelin components possibly enhanced and the
capacity for resynthesis decreased.

The manner in which this large area of highly specialized
membrane is maintained by the relatively small soma of the
oligodendrocyte or Schwann cell presents an interesting problem in cell biology, perhaps somewhat similar to the maintenance of a long axon by a neuron. Perhaps glial cells have
as yet uncharacterized specialized systems for membrane insertion and transport. The liquid crystalline structure of the
membrane which allows diffusion of protein and lipid constituents along its length and thus through the myelin sheath is
undoubtedly important. Possibly the portions of cytoplasm in
the Schmidt-Lantermann incisure and paranodal regions are also
involved in the grooming and maintenance of myelin.

Questions:
1) What further protein and lipid constituents of myelin remain to be discovered?
2) Where is myelin basic protein located in the membrane?
3) How do glial cells maintain their membrane expansions?
How do cytoplasmic extensions and other cell organelles move
through the layers of the myelin sheath?
4) Can viral coat proteins be incorporated into myelin?
5) Does the surface of the oligodendrocyte soma as opposed
to myelin possess different antigenic markers, surface constituents, viral receptors? Could these be important in the
pathogenesis of multiple sclerosis?

MAINTENANCE OF THE EXTRACELLULAR ENVIRONMENT - IONS

The amount of extracellular space in the CNS is sufficiently
small that the ionic composition of its fluid can be altered
significantly by the ion fluxes accompanying normal neuronal
electrical activity. Recordings with ion-sensitive electrodes
have shown that during intense neural activity the concentration of extracellular K^+ can increase five-fold, while extracellular Ca^{2+} may decrease by one third (Orkand, this volume).
Such shifts in the ionic milieu are likely to alter neuronal
physiology, particularly transmission of trains of repetitive
stimuli and release of synaptic transmitters.

Glial cells have been suggested to play a role in maintaining
the normal ionic composition of extracellular fluid, particularly in K^+ homeostasis. They could help regulate extracellular K^+ by either serving as a spatial buffering system or by
actively transporting K^+ into their cytoplasm. Observations
on the electrophysiological properties of glial cells provide
evidence consistent with the first mechanism. Recordings from
glial cells in the optic nerve of Necturus demonstrate that
the cell behaves like a K^+-sensitive electrode, indicating a
selective permeability to K^+. During neuronal activity, the
glial cell becomes depolarized by K^+ released by the neurons
(Orkand, this volume). K^+ could be driven into the glial cell
at the site of depolarization by the electrochemical gradient
across the membrane, then distributed to other glial cells in
less active areas through connecting gap junctions, and finally
extruded. The current loop would be completed by movement of
Na^+ and Cl^- in the extracellular space. The electrical space
constant of the Necturus glia is about 0.8 mm (Tang, unpublished), which is sufficiently long to allow significant spatial redistribution of K^+ ions. K^+ has been demonstrated to
move through the glia of the insect retina (24), but the unusually high potassium concentration of blood plasma in in-
sects poses different problems for long-term ionic homeostasis
of the brain microenvironment (90) than in mammals. More direct observation of K^+ movements in mammalian glia would clearly be of interest.

Glia do have a strophanthidin-sensitive Na^+/K^+ ATPase which
could function in active transport of K^+ ions out of the extracellular space. In other cell types, however, this ion pump
is activated by changes in the concentration of intracellular
Na^+ rather than extracellular K^+. Nonelectrogenic pumping
mechanisms, like a KCl pump, are also possible, though there
is little data to support their existence. The major role of
glia in maintaining extracellular K^+ concentration, therefore,
would probably be as a spatial buffer (73).

Although the spatial buffering scheme is a plausible one,
there are many quantitative considerations which require further consideration. For example, since the membrane resistance of glial cells is quite high, it is unclear whether they
can allow sufficient K^+ influx to significantly effect the extracellular concentration of K^+. On the other hand, the total
quantity of K^+ moving during neuronal activity is small, on the
order of $pmoles/cm^2$ neuronal membrane (Orkand, this volume). Attempting to apply this model to the normally functioning brain also
raises difficulties, since neuronal activity occurring simultaneously in adjoining areas would seem likely to set up many competing
and conflicting currents within the glial syncytium. Further
experimental observation and computer modelling should be able
to settle these questions.

The significance of these postulated ion movements for both
the neurons and the glial cells are uncertain. From the perspective of the neuron, lowering extracellular K^+ during
activity could help sustain neuronal activity and transmitter
release in the short run. In the long run, however, it might
retard the reuptake of K^+ and attainment of the normal ionic
composition of the resting state. For the glia, increases in
extracellular K^+ have been demonstrated to decrease the level
of NADH, and to increase glucose uptake and incorporation into
organic and amino acids (Orkand, this volume). It is unclear, however, if increased glial uptake of glucose is productive for the
neurons, since neuronal energy use increases during activity and

they are competing for the same glucose supply. In a number
of cell types, large changes in cellular K^+ have been observed
to influence many metabolic processes, including protein and
DNA synthesis (62, 68). It is unknown whether such effects
might occur in glial cells in response to the K^+ fluxes accompanying neuronal activity.

While the role of glia in brain K^+ homeostasis is still unclear, there is so little data concerning regulation of the
other ionic constituents as to make the situation nearly speculative. There are a few interesting observations, however,
which may bear on regulation of some of these other ionic
species. For example, astrocytoma cells are known to have
significant Cl^- flux rates (58). The enzyme carbonic anhydrase appears to be specifically localized to oligodendrocytes in the CNS (37). In other nonneuronal cells, carbonic
anhydrase appears to aid in the transport of Cl^- or H^+ ions by
catalyzing the conversion of CO_2 to HCO_3^- (55). These observations together could suggest the existence of mechanisms for
controlling the pH and anion contents of the extracellular
fluid.

In neuronal tissues, control of extracellular Ca^{++} concentration is particularly important because of its influences on
membrane threshold, axonal conduction block (44) and the
release of neurotransmitter. Intracellularly, Ca^{++} in con-
junction with its binding protein, calmodulin, also appears to
regulate a number of metabolic pathways (21, 69). Microprobe
analysis of brain tissue provides evidence for stores of Ca^{++}
within astrocytes (36). Electron microscopic observation of
neuronal submembranous cisternae under astrocyte processes also suggests a possible Ca^{++} transfer (Mugnaini, this volume).
Further analysis of the homeostasis and regulation of Ca^{++} in
the CNS are clearly of interest.

Questions:
1) In addition to K^+, might glial cells help regulate other
extracellular ions: Ca^{2+}, H^+, Cl^-, HCO_3^-?

2) Does the glial depolarization during neuronal activity affect glial metabolism or growth?
3) Do alterations in the neuronal ionic microenvironment because of glial disease further affect neuronal physiology in
demyelination and multiple sclerosis?
4) Since viral replication can be effected by intracellular
ion concentrations, do glial ion fluxes during neuronal activity have a role in the pathophysiology of CNS viral infections,
particularly multiple sclerosis?

MAINTENANCE OF TRANSMITTER METABOLISM

The concentration of transmitter substances in the extracellular space must be closely regulated to prevent desensitization
of receptor molecules and allow normal synaptic response.
Glial cells may aid in the inactivation of some synaptic transmitters, by uptake and metabolism. This has been particularly
well characterized for GABA in the satellite cells of mammalian sympathetic (12) and sensory (85) ganglia. The uptake
system is fairly specific for GABA, with an apparent transport
constant of 6.8 µM. Using the energy provided by the Na^+-
gradient, glial cells can accumulate millimolar concentrations
of GABA intracellularly when micromolar concentrations are
present extracellularly (14). GABA uptake has also been demonstrated into the cytoplasm of myelinating Schwann cells in
peripheral nerve (101), and into CNS oligodendrocytes and astrocytes (53) by autoradiography. The glial cells can trans-
aminate GABA and process the products through the Krebs cycle
(96), so eliminating the transmitter. The glial uptake
system for GABA has a relatively low velocity when compared to
the GABA-uptake system found in terminals of GABAergic neurons,
but a much higher capacity. At synaptic sites, the role of
the glial pump is likely secondary to that of the neuron, but
it probably functions in inactivating transmitter diffusing
away from synaptic sites and in maintaining a low background
level of GABA in the CSF. Similarly, glial cells have been
shown to take up glutamate (84) and, in the spinal cord, glycine (48).

In addition to uptake systems, there is some evidence that
glial cells may possess receptors for neurotransmitter compounds. The glial cells of Necturus respond to administration
of 10^{-5}M glutamate with depolarization. This clearly does not
represent a secondary response to K^+ release by depolarized
axons, since it continues after degeneration of the axons
(Orkand, this volume). Na^+ influx appears to be involved
(Orkand, this volume), though whether due to combined Na^+ and
glutamate uptake (52) or to the opening of Na^+ channels by
specific receptors is unclear. Villegas has reported that
acetylcholine hyperpolarizes the Schwann cells of squid axon
by interacting with a bungarotoxin-sensitive receptor (94),
though these results have not yet been confirmed by independent investigators. In cultures of glial cells, exposure to
catecholamines has been reported to increase cAMP and enhance
glycogenolysis (93), though there is no evidence for electrophysiological receptor for catecholamines.

Reports of synaptic profiles from neurons ending on glial
cells further suggest the possibility of neuronal-glial communication using neurotransmitter compounds. Such synapses
have been observed in developing embryonic tissue (77, 100)
and in adult tissue onto ependymal cells (63) and astrocytes
in periventricular regions (28, 45, 61). The relative
paucity of these reports raises the question of whether such
synapses are exceptional, or have simply received little attention. Even without morphological synapses, glial cells
could still be affected by neurotransmitter released at vari-
cosities (26, 27) or diffusing away from synaptic clefts.

Conversely, it is also possible that glial cells could release
neurotransmitter compounds onto neurons. In addition to
taking up GABA, astrocytes and Schwann cells appear to contain
glutamic acid decarboxylase and to be able to synthesize GABA.
Though this ability is small compared to that of neurons, it
may be significant, with sympathetic ganglia containing about
1/5 of the GABA synthesizing capacity of a comparable quantity

of cortical tissue (51). Strong depolarizing stimuli can enhance the release of GABA from glial cells (13), though this
process does not have the Ca^{++} sensitivity expected of vesicular release. Together these observations raise the possibility that glia depolarized by K^{+} released from active neurons
may release GABA to feed back on the neurons. However, some
quantitative observations argue against this idea. The concentration of K^{+} needed to stimulate GABA release is quite
high, with threshold on the order of 20-50 mM, and the quantities of GABA released are quite low (13). It appears that decreased activity or even reversal of the normal GABA uptake
mechanism with changes in the membrane potential may be responsible for the apparent GABA release. Experiments in which
peripheral nerve glia were loaded with GABA, and then the
axons stimulated with trains of impulses, failed to show significant GABA release (13). This leaves open the question of
the function, if any, of peripheral GABA receptors or of GABA
synthesized by glial cells.

Schwann cells at the denervated neuromuscular junction of
frogs have been observed to release acetylcholine (10), and to
contain choline acetyl transferase (91). The presence of ac-
tinomycin-D during terminal degeneration prevents acetylcholine release at later times (10), suggesting that cholinergic
function could be induced in the Schwann cells. The function
of the acetylcholine produced is uncertain, though trophic
effects on the denervated muscle have been suggested.

Questions:
1) How frequently do synapses between neurons and glia occur?
What, if any, is their functional significance?
2) Do glial cells possess receptors for a variety of neurotransmitters? Do these effect glial metabolism and function?
3) Is glial neurotransmitter synthesis induced or increased
in other pathological states? What is its significance?
4) Are there changes of extracellular concentrations of neurotransmitters in multiple sclerosis? Could these be related to
observed symptoms?

MAINTENANCE OF METABOLIC INTERMEDIATES

It has long been suggested that a major function of glial cells
may be to control the availability of small molecules of metabolic importance to neurons, such as glucose and amino acids.
One example of metabolic interdependence is a postulated glutamine cycle in which glia take up glutamate released by neurons,
convert it to glutamine, and return this amino acid to the
neurons (5). In support of this hypothesized cycle are observations on the different labelling kinetics of pools of glutamate
and glutamine in bulk isolations of neuronal and glial tissues
from whole brain. There is also evidence for glutamate carrier
systems on glial cells and glutamine systems on neurons. Biochemical studies indicate that glutamine synthetase is restricted to glial cells, and immunohistochemical studies further
define its localization in astrocytes (66). The functional
significance of this cycle is uncertain, though there are suggestions that neurons may use glutamine as a preferred energy
source under some conditions (59). The existence of this
interesting neuronal-glial cycle is still only inferential and
requires more direct demonstration.

Question:

1) Do neurons suffer from a shortage of metabolic intermediates
(sugars, amino acids) in pathological states affecting the glia?
Could exogenous supply of these compounds improve neuronal function?

MAINTENANCE OF EXTRACELLULAR MATRIX MATERIALS

The Schwann cell related to nerve fibers is surrounded by a
basal lamina which separates it from the remaining endoneural
components. Observations on Schwann cells and neurons grown
without accompanying fibroblasts in tissue culture demonstrate
that at least a portion of this basal lamina can be formed.
Some small collagen fibrils were also formed, which on biochemical analysis included types I, III, V, and possibly IV (18).
In addition to this indication of collagen production by Schwann

cells, there is evidence that Schwann cell contact with extracellular matrix materials (presumably of fibroblast origin) is
required during early stages of Schwann cell differentiation
(Bunge, this volume).

There is no morphologically demonstrable basal lamina within the
CNS itself, though some flocculant material is occasionally seen
in expansions of the extracellular space.   Glycosaminoglycans
are found within the brain, however, in both soluble and particulate form.   About half of the hyaluronic acid and heparan
sulfate, and about one-third of the chondroitin sulfate are
found in microsomal fractions, presumably attached to plasma
membrane (54).   In cell culture, glial-like cells and gliomas
have been observed to synthesize all three types of glycosaminoglycans (39).

The role of these glycosaminoglycans in the brain is speculative, though they have been suggested to be structural components of the intercellular space, or to play a role in morphogenesis (65, 99).   Such extracellular molecules could also have an
important influence on ionic homeostasis.   In the insect CNS
where hyaluronic acid is a major extracellular constituent (4),
it has been suggested to serve an important functional role in
binding K^+ ions released during neuronal activity and in releasing Na^+ associated with its anionic groups (89).

Questions:
1)   What is the role of glycosaminoglycans in the CNS?
2)   Are other large molecules present in the extracellular space?

MAINTENANCE OF BARRIERS
The barrier which restricts passage of substances between the
blood and the brain substance is well recognized.   Another barrier exists in the periphery which restricts the passage of
substances from the blood into the endoneurium of nerves.   This
peripheral blood-nerve barrier appears to be absent in peripheral and enteric ganglia, allowing free passage of materials

from the blood to the extracellular space surrounding the neuronal somata (50). The existence of this peripheral blood-nerve
barrier is based mainly on observations of horseradish peroxidase diffusion (76), and it is not clear whether it presents a
similar limitation to the movements of ions and small molecules
as the blood-brain barrier.

Although it has long been debated whether the astroglial processes lining intracerebral blood vessels contribute to the
blood-brain barrier, the major portion clearly lies in the
tight junctions formed between capillary endothelial cells (16).
Myelinating glial cells do appear to also form barriers
dividing up the extracellular space. At each node of Ranvier,
the myelinating cell and its axon form a series of unique junctions which bear some morphological features of tight junctions.
These junctions appear to seal off the extracellular space
immediately overlying the axon, although a small slow penetration of microperoxidase into this space has been seen (Mugnaini,
this volume).

MAINTENANCE OF MEMBRANES

The membranes of neurons and their myelinating glial cells appear to be able to exchange and directly share their lipid constituents. After labelled glycerol is injected into chicken
brain near the oculomotor nucleus, labelled lipid constituents
are transported along the axons of this nerve (29). Within 3-6
hours label can be detected autoradiographically in the myelin
lamellae of the Schwann cell. Label first appears mainly in the
Schmidt-Lantermann incisures and in the innermost layers of the
myelin lamellae. With time, it spreads throughout the thickness
of the Schwann cell, though a gradient with greatest concentration near the axon is maintained. Labelling is heavier over the
internode than near the juxtanodular specializations. About 60%
of the glycerol ends up in phosphatidylcholine, with the remainder in a variety of other lipids. This lipid does not appear to
be synthesized from isotope leaked to the glia because systemically injected labelled glycerol is not similarly used by the

Schwann cell. Similar movement of labelled lipids synthesized
from choline has also been seen, though the final pattern of distribution in the Schwann cell is slightly different (Fig. 1). Transfer of lipids between axons and CNS myelin in the optic tract
has also been observed (46). The observations are consistent
with the interpretation that lipid is being transferred from the
neuronal membrane to the glial membrane, perhaps by carriers
located in the extracellular space between the internodal axon
and myelin. Such an equilibration mechanism is hypothetical.
The transferred lipid could serve a role in the maintenance of
the myelin lamellae of the innermost region of the sheath. The
transfer of lipid from the axon to the inner myelin sheath is
considerably quicker than the movement of newly synthesized
lipids from the glial cell body. The amount of transferred
lipid appears to be small, however, raising the question of its
significance in maintaining the myelin.

The maintenance of localized spatial differences in structure is
one of the more interesting properties of membranes and likely
to be influenced by interactions with other cell types. For
example, myelin-associated glycoprotein (MAG) is not evenly
distributed throughout myelin but is found concentrated in the
innermost portion near the axon (49). Since MAG is suspected to
bind to an axonal component during initiation of myelination,
this inhomogeneity could be related to the presence of the axon.

The most striking inhomogeneity of the axonal membrane relates
to the distribution of the ionic channels. Under normal circumstances, the nodal membrane contains a dense population of voltage sensitive Na^+ channels, which the internodal membrane apparently lacks. After demyelination, axons with diffusely distributed Na^+ channels and continuous conduction have been observed (Bostock and McDonald, this volume). It seems likely
that the interaction of the myelinating glial cell and neuron
is important in maintaining this complex spatial distribution.
Whether the neuron or the Schwann cell plays the directive role

186   State of the Art Report; P.A. Walicke et al.

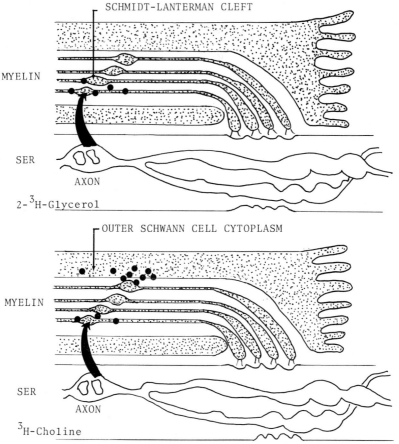

FIG. 1 - Axon-glial transfer of phospholipid components.
In the course of the axonal transport of phospholipids labeled
with 2-^{3}H-glycerol, radioautography with the electron microscope
shows an axon-myelin transfer of radioactive phospholipids, especially in the Schmidt-Lanterman clefts and the inner myelin
lamella (black dots).

When the axonally transported phospholipids are labeled with
^{3}H-choline, a similar axon-myelin transfer of radioactive phospholipids is observed. In addition, the labeling of the outer
Schwann cell cytoplasm and of the outer myelin-lamellae indicates that free choline of axonal origin is reincorporated into
new phospholipids, mainly in the vicinity of the Schmidt-Lanterman
clefts.

is uncertain, though some evidence from studies of remyelination supports a primary role for the axon. For example, electrophysiological recordings along the length of a demyelinated axon demonstrate points of increased excitability before remyelination (Bostock and McDonald, this volume). In a similar situation at the neuromuscular junction, however, molecular determinants retained in the basal lamina appear to play a role in inducing acetylcholine receptor aggregation and specifying endplate location in the muscle cell (20). An analogous signal effecting Na^+ channel distribution may be located in the basal lamina left behind by the Schwann cells which overlies the demyelinated axons. At this point little can be concluded about the cellular signals allowing maintenance of Na^+ channel distribution, although the possible roles of axon, Schwann cell, and basal lamina all deserve further investigation.

Questions:
1) How significant is the amount of lipid transferred between cell types to both neuronal and glial metabolism? Does loss of myelination affect axonal lipid constitution?
2) Is the distribution of Na^{++} channels in the axon determined by the axon, the myelinating glial cells, or by a molecular signal deposited in the basal lamina?

MAINTENANCE OF THE CYTOSKELETAL NETWORK
Cytoskeletal proteins serve in a variety of cellular functions including cell motility, intracellular transport, exocytosis, maintenance of distribution of membrane elements, and possibly in transduction of transmembrane signals. Perhaps their most basic function is in conferring a stable shape on the cytoplasm. Axonal caliber appears to correlate most closely with neurofilament number/cross sectional area (33, 75), and neurofilaments have been suggested to be a major intrinsic determinant of axon size. These filaments are arranged in fascicles, with a relatively fixed interfascicular distance (97), and their most important known function is this ability to constructively take up space. The relationship between microtubule number or

other cellular organelle number and cross sectional area is
weak. Since neurofilaments undergo slow axonal transport, the
role of slow transport in regulation of fiber caliber also
deserves attention.

Myelinated axons in the peripheral nervous system are not constant in diameter along their length. At the nodes and at the
site of myelin sheath attachment, their diameter is as much as
70-80% smaller than in the internode (Mugnaini, this volume).
Serial electron micrographs have shown that the number of microtubules is nearly constant between internode and node, while the number of neurofilaments is greatly reduced (9). The mechanism by
which neurofilaments traverse the node of Ranvier is unknown. Possibly they simply move through the node at a greater velocity
than through the internode. Since Ca^{++} concentration affects
neurofilament polymerization (38), it is possible that the
higher Ca^{++} concentration in the node causes reversible depolymerization with repolymerization in the internode. Free neurofilament subunits, however, have not been observed in normal
neuronal tissue. It is also unclear whether the axon or the
myelinating glial cell is responsible for the initiation and
maintenance of the constricted segments.

It has long been recognized that demyelination leads to a substantial local decrease in axonal diameter. This has been
observed in experimental allergic neuritis (80), after lysolecithin administration (47), and in the trembler mouse mutant (2).
The reduction in axonal diameter persists for the duration of
demyelination. Quantitative electron microscopy to determine
whether this change in axonal caliber is due to a decrease in
neurofilament number has not yet been done. It does appear
possible though that myelinating glial cells can influence
axonal size and maybe even the distribution of intracellular
neurofilaments.

Questions:
1) What is the significance of the change in axon caliber after
demyelination in terms of electrophysiology, axonal survival?

2) Why are nodes of Ranvier or demyelinated axons smaller? Is
a signal from the glial cells involved?

MAINTENANCE OF OTHER PROTEIN CONSTITUENTS
The possibility that intact proteins could pass between glial
and neuronal cytoplasms has often been suggested. Transfer
of proteins synthesized by glial cells into axons does appear
to occur, at least in invertebrates. If the squid axon is can-
nulated and labelled amino acids supplied to its satellite
cells, labelled proteins appear in the axonal perfusate. This
observation does not appear to be due to neuronal synthesis, because amino acids directly injected into the axon itself are not
incorporated into protein (60). This ability of the glial cells
to directly supply axonal proteins may explain the ability of
axons severed from cell bodies to survive for many months in invertebrates (11).

The situation in the vertebrate neuron is less clear. Administration of labelled histidine to cultured dorsal root ganglion
neurons has been observed to result in the appearance of label
detected by autoradiography over the axon at relatively short
times (86). There is no evidence, however, to indicate whether
this label is in proteins or free amino acids. More thorough
examination of this important question in vertebrate nerves is
clearly needed. Even if passage of entire proteins is demonstrated, it will still be necessary to show that the proteins
have functional significance. Most cell types appear to periodically take up pinocytotic or endocytotic vesicles for uncertain
reasons, perhaps just to monitor the extracellular environment.
Much of the content of these vesicles quickly ends up in lyso-
somes. Therefore, it is necessary to demonstrate that transferred
proteins serve a function and are not simply an accidental in-
clusion in this random endocytosis.

Questions
1) Does functionally significant transfer of proteins from glial
cells to axons occur? Could their absence affect axonal survival
in demyelinating diseases like multiple sclerosis?

2)   If proteins are transferred from glia to neurons, do they
include growth factors?

MAINTENANCE OF MYELIN AND THE GLIAL CELL
The thickness of the myelin sheath and the length of the internode are both correlated with the diameter of the ensheathed
axon.   Evidence has already been presented that myelination in
turn affects axonal diameter (see Maintenance of Cytoskeleton).
Gradual changes in axonal diameter during growth occur long after establishment of myelination and must be balanced by expansion and increase in the thickness of the myelin sheath.   In
many pathological situations, axonal caliber can be altered, and,
at least initially, compensatory changes in the glial cell seem
to be accommodated by simple physical deformation.   Axonal swelling results in thinning of the myelin sheath (32, 33), perhaps
reflecting slippage of myelin lamellae over each other.   Shrinkage of the axon results in a myelin sheath disproportionately
thick for axonal caliber, with a "crenated" or "knuckled" appearance in cross-section (22, 30).   It has been suggested that
after prolonged reduction in axonal diameter, more definitive
restructuring of the myelin sheath may be accomplished by segmental demyelination and remyelination at an appropriate size
and thickness (30).   The possible existence of such secondary
demyelination occurring in response to primary axonal disease
can be studied critically with currently available techniques
(2, 43).   The very survival of glial cells seems to be at least
partially dependent on the presence of the axon.   After prolonged denervation, glial cells are lost from the distal stump
of the transected tract in both the peripheral and central
nervous systems ((34), and Orkand, this volume).

These observations indicate that some sort of signals must pass
from axon to myelinating cell in the adult animal to insure
glial survival and guide modelling of the myelin sheath.   The
time course of these adjustments in the adult is relatively
slow compared to that during development, which could indicate
a difference in the type or availability of the axonal signal,
or in the dependence and responsiveness of the glial cell.   It
is not unlikely that the signals operating during normal myelin

maintenance are the same as those during development or remyelination. At this point in time, development and remyelination have also been more thoroughly studied, and the suspected signals provide at least a framework for speculation
on maintenance of the normal state.

It has been hypothesized that direct interactions between
myelin-associated glycoprotein (MAG) and axonal surface proteins might occur (49), possibly these could be involved in
some of the physical changes in myelin sheath accompanying axonal swelling or shrinkage. The axonal surface agent responsible for providing a mitogenic signal to the Schwann cell
during early development ((82), and Bunge, this volume) must
also be masked or altered as Schwann cells ensheathe axons and
cease proliferation. This masking may be mediated by some secreted product of the Schwann cell, because when secretion is
blocked, Schwann cell proliferation continues at abnormally high
levels ((72), and Bunge, this volume). Observations on tissue
cultured Schwann cells and nerves in the absence of fibroblasts
indicate that original expression of myelination is dependent
on extracellular signals which can be provided by addition of
embryo extract or serum (72). This factor or factors does not
appear to be required for maintenance of the myelinated state
in tissue culture.

A variety of growth factors are recognized to be able to influence glial cells and could be involved in maintaining glial
survival and differentiation. These include glial growth factor (GGF) and fibroblast growth factor (FGF) from bovine pituitary and brain, and epidermal growth factor (EGF) from mouse
salivary gland. GGF stimulates mitosis of Schwann cells and
astrocytes (17). FGF stimulates the proliferation of a variety
of mesoderm-derived cells (40) and affects the mitosis of human
glial cells having the morphology of astrocytes (98). A mitogenic response of GFA-positive cells to FGF has recently been
described (R. Mirsky, personal communication). EGF is a
mitogen for certain epidermal cells in vivo (83) and for

a variety of cells in vitro (23, 41).   EGF stimulates the
proliferation of oligodendrocytes during a certain   period
of development (K. Herrup,   E. Trenckner, personal communication).   Studies using growth factors thus far have concentrated on their ability to stimulate mitosis, rather than
their possible roles in influencing survival and differentiation.
FGF has been reported to effect the differentiation of other
cell types   (95), though not of astrocytes or oligodendrocytes.
To determine whether any of these factors or the axolemmal mitotic signal are related to physiological maintenance of mature
glial cells, a number of questions must be addressed, including:
a) effects on glial cell survival, b) effects on glial cell
growth, c) effects on expression of differentiated glial functions, d) demonstration of its presence and distribution in the
normal adult.   It is possible that one or more of these factors
may have pleiotropic effects on glial cell survival and differentiation, as well as influencing mitosis during development or
remyelination.   Conversely, a separate spectrum of "differentiation factors" acting on glial cells may yet be awaiting discovery.

Schwann cells and oligodendrocytes seem to differ in their
ability to maintain the differentiated state for myelin production.   Oligodendrocytes can be grown in tissue culture in the
absence of neuronal cells, though in the presence of astrocytes,
where they continue to express galactocerebroside, basic protein and other specific myelin markers (70).   Schwann cells
grown alone in culture, however, lose the expression of galactocerebroside and other myelin markers over several days.   These
observations are very interesting though open to many explanations as yet, including differences in the signals affecting
Schwann cell and oligodendrocyte differentiation, differences in
the dependence of these cell types on a similar neuronal signal,
or a possible role of astrocytes in oligodendrocyte differentiation.

Questions:
1) What is the specific stimulus for correlation between axonal

caliber and myelin parameters?
2)   Do any of the recognized mitotic signals also influence glial
survival, growth, and differentiation?
3)   How frequently does "secondary demyelination" in response
to axonal disease occur?
4)   Do proteases and other products released by macrophages
phagocytizing myelin stimulate mitosis or growth of myelinating
glial cells or astrocytes?

MAINTENANCE OF THE NEURON

It has been postulated that the survival of neuronal cells is
dependent on the provision of specific protein growth factors.
Glial cells may be able to synthesize these agents and, thus,
play a role in maintaining the neuron itself. The only one of
these factors which has been isolated and characterized to
date is nerve growth factor (NGF), which is required for the
survival of sympathetic neurons and a subpopulation of dorsal
root ganglion cells. Homologous non-neuronal cells from the
ganglia can support the survival of the neurons in vitro (92),
and presumably NGF may be involved in this effect. A factor
functionally and immunologically similar to NGF also appears
to be produced by astrocytes (64) and, at low levels, by C6
glioma cells (6).

Recently, glial production of neuronal growth factors besides
NGF has been described. C6 glioma cells produce a factor which
can be distinguished functionally from NGF on the basis of the
age of responsive sensory neurons (6, 7). Similar activity can
also be detected in extracts of rat brain (6). Astrocytes in
culture produce diffusible factor(s) which can support NGF-
independent neurons like those of the nodose ganglion and the
unresponsive subpopulation of the dorsal root ganglion (64).
Although these factors have not been purified, they do not
cross-react with antibody to NGF (6, 7, 64).

The physiological role of these factors requires more direct
demonstration, including observation of their production in vivo
and examination of the effects of injection of specific antibodies in animals of various ages. Available evidence already
indicates that these factors may act at different times ("windows")

during development, and that a single neuronal population may
respond to more than one factor (6). Target organs also appear
to produce growth factors (1, 31), and the relative importance
of glial and target organ supplied factors needs to be assessed.

In addition to growth factors, glial cells may be able to
produce "differentiation signals" which can affect the expression of specific neuronal properties. For example, sympathetic
neurons grown in tissue culture in the presence of fibroblasts and
ganglion satellite cells are induced to change from adrenergic to
cholinergic differentiation. A specific protein factor appears to
be responsible for this transformation and is clearly distinct
from the growth factors previously discussed in that it does
not influence neuron survival, protein synthesis, or lipid content (78). The possible existence of other such transmitter-
specifying differentiation signals and their neuronal role in
vivo is clearly of interest.

Questions:

1) Do glial cells make NGF and other neuron-specific growth
factors?

2) Do glia play a significant role in guiding neuronal differentiation through factors like the cholinergic induction
factor?

3) Why do axotomized neurons survive longer in the CNS than
in the periphery? Could this reflect provision of necessary
survival factors by surrounding glial cells?

4) Does axonal loss in chronic demyelinated states like
multiple sclerosis also reflect loss of growth and survival
factors normally provided by the glial cell?

REFERENCES

(1) Adler, R.; Landa, K.B.; Manthorpe, M.; and Varon, S. 1979.
     Cholinergic neuronotrophic factors: Intraocular distribu-
     tion of trophic activity for ciliary neurons. Science 204:
     1434-1436.

(2) Aguayo, A.; Bray, G.M.; and Perkins, S.C. 1979. Axon-
     Schwann cell relationships in neuropathies of mutant mice.
     Ann. NY Acad. Sci. 317: 512-531.

(3) Aguayo, A.J.; Charron, L.; and Bray, G.M. 1976. Poten-
     tial of Schwann cells from unmyelinated nerves to produce
     myelin: a quantitative ultrastructural and radiographic
     study. J. Neurocytol. 5: 565-573.

(4) Ashhurst, D.E., and Coston, N.M. 1971. Insect mucosub-
     stances. III. Some mucosubstances of the nervous system
     of the wax moth (Galleria mellonella) and the stick insect
     (Carausius morosus). Histochem. J. 3: 379-387.

(5) Balazs, R., and Cremer, J.E. 1973. Metabolic Compartmenta-
     tion in the Brain. London: MacMillan.

(6) Barde, Y.A.; Edgar, D.; and Thoenen, H. 1980. Sensory neu-
     rons in culture: Changing requirements for survival factors
     during embryonic development. Proc. Natl. Acad. Sci. 77:
     1199-2203.

(7) Barde, Y.A.; Lindsay, R.M.; Monard, D.; and Thoenen, H.
     1978. New factor released by cultured glioma cells sup-
     porting survival and growth of sensory neurones. Nature
     274: 818.

(8) Benjamins, J.A., and Smith, M.E. 1977. Metabolism of
     myelin. In Myelin , ed. P. Morell, pp. 230-270. New York:
     Plenum Press.

(9) Berthold, C.-H. 1968. Ultrastructure of the node-paranode
     region of mature feline ventral lumbar spinal root fibers.
     Acta Societatis Medicorum Uppasaliensis 73, Suppl. 9: 37-70.

(10) Bevan, S.R.; Miledi, R.; and Grampp, W. 1973. Induced
     transmitter release by Schwann cells and its suppression by
     actinomycin D. Nature, New Biol. 241: 85-86.

(11) Bittner, G.D.; Ballinger, M.L.; and Larimer, J.L. 1974.
     Crayfish CNS: Minimal degenerative-regenerative changes
     after lesioning. J. Exp. Zool. 189: 13-96.

(12) Bowery, N.G., and Brown, D.A. 1972. γ-Aminobutyric acid
     uptake by sympathetic ganglia. Nature, New Biol. 238: 89-91.

(13) Bowery, N.G.; Brown, D.A.; and Marsh, S. 1979. γ-Amino-
     butyric acid efflux from sympathetic glial cells: Effect
     of "depolarizing" agents. J. Physiol. 293: 75-101.

(14) Bowery, N.G.; Brown, D.A.; White, R.D.; and Yamini, G.
     1979. (^{3}H)-γ-aminobutyric acid uptake into neuroglial
     cells of rat superior cervical sympathetic ganglia. J.
     Physiol. 293: 51-74.

(15) Braun, P.E. 1977. Molecular architecture of myelin. In
     Myelin, ed. P. Morell, pp. 91-115. New York: Plenum Press.

(16)   Brightman, M.; Vlatzo, I.; Olsson, Y.; and Reese, T.   1970.
       The blood brain barrier to proteins under normal and patho-
       logical conditions. J. Neurol. Sci. $\underline{10}$: 215-239.

(17)   Brockes, J.P., Lemke, D.R.; Balzer, J.R.   1980. Purifica-
       tion and preliminary characterization of a glial growth
       factor from the bovine pituitary. J. Biol. Chem. $\underline{255}$:
       8374-8877.

(18)   Bunge, M.B.; Williams, A.K.; Wood, P.M.; Uitto, J.; and
       Jeffrey, J.J.   1980. Comparison of nerve cell and nerve
       cell plus Schwann cell cultures, with particular emphasis
       on basal lamina and collagen formation. J. Cell Biol. $\underline{84}$:
       184-202.

(19)   Bunge, R. 1968. Glial cells and the central myelin sheath.
       Physiol. Rev. $\underline{48}$: 197-251.

(20)   Burden, S.J.; Sargent, P.B.; and McMahan, U.J. 1979.
       Acetylcholine receptors in regenerating muscle accumulate
       at original synaptic sites in the absence of the nerve.
       J. Cell. Biol. $\underline{82}$: 412-425.

(21)   Cheung, W.Y. 1980. Calmodulin plays a pivotal role in
       cell regulation. Science $\underline{207}$: 19-27.

(22)   Clark, A.W.; Griffin, J.W.; and Price, D.L. 1980. The
       axonal pathology in chronic IDPN intoxication. J. Neuro-
       path. Exper. Neurol. $\underline{39}$: 42-55.

(23)   Cohen, S.; Carpenter, G.; and Lembach, K.J. 1975. Inter-
       action of epidermal growth factor (EGF) with cultured fi-
       broblasts. Adv. Metabl. Disorders $\underline{8}$: 265-284.

(24)   Coles, J.A., and Tsacopoulos, M. 1979. Potassium activ-
       ity in photoreceptors, glial cells and extracellular spaces
       in drone retina: Changes during photostimulation. J.
       Physiol. $\underline{290}$: 525-549.

(25)   Davison, A.N., and Cuzner, M. 1977. Immunochemistry and
       biochemistry of myelin. Br. Med. Bull. $\underline{33}$: 60-66.

(26)   Descarries, L.; Beaudet, A.; and Watkins, K.C. 1975.
       Serotonin nerve terminals in adult rat neocortex. Brain
       Research $\underline{100}$: 563-588.

(27)   Descarries, L.; Watkins, K.C.; and Lapierre, Y. 1977.
       Noradrenergic axon terminals in the cerebral cortex of
       rat. III. Topometric ultrastructural analysis. Brain
       Research $\underline{133}$: 197-222.

(28)   Diederen, J.H. 1970. The subcommisural organ of Rana
       temporaria. I. A cytological, cytochemical, cytoenzymo-
       logical and electron microscopical study. Z. Zellforsch.
       $\underline{111}$: 379-403.

(29) Droz, B.; Di Giamberardino, L.; Koenig, N.J.; Boyenval, J.; and Hassig, R. 1978. Axon-myelin transfer of phospholipid components in the course of their axonal transport as visualized by radioautography. Brain Research 55: 347-353.

(30) Dyck, P.J. 1975. Pathologic alterations of the peripheral nervous system of man. In Peripheral Neuropathy, eds. P.J. Dyck, P.K. Thomas, and E.H. Lambert, p. 296. Philadelphia: W.B. Saunders.

(31) Ebendal, T.; Olson, L.; Seiger, Å.; and Hedlund, K.O. 1980. Nerve growth factors in the rat iris. Nature 286: 25-28.

(32) Friede, R.L., and Miyagashi, T. 1972. Adjustment of the myelin sheath to changes in axonal caliber. Anatomical Record 172: 1-14.

(33) Friede, R.L., and Samorajski, T. 1970. Axon caliber related to neurofilaments and microtubules in sciatic nerve fibers of rats and mice. Anatomical Record 167: 379-388.

(34) Fulcrande, J., and Privat, A. 1977. Neuroglial reactions secondary to Wallerian degeneration in the optic nerve of the postnatal rat: Ultrastructural and quantitative study. J. Comp. Neurol. 176: 189-224.

(35) Futamachi, K., and Pedley, T.A. 1977. Glial cells and potassium: Their relationship in mammalian cortex. Brain Res. 109: 311-322.

(36) Gambetti, P.; Erulkar, S.; Somlyo, A.P.; and Gonatas, N.K. 1975. Calcium-containing structures in vertebrate glial cells, ultrastructural and microprobe analysis. J. Cell Biol. 64: 322-330.

(37) Giacobini, E. 1962. A cytochemical study of the localisation of carbonic anhydrase in the nervous system. J. Neurochem. 9: 169-177.

(38) Gilbert, D.A.; Newby, B.J.; and Anderton, B.H. 1975. Neurofilament disguise, destruction and discipline. Nature 256: 586-589.

(39) Glimelius, B.; Norling, B.; Westermark, B.; and Wasteson, Å. 1978. Composition and distribution of glycosaminoglycans in cultures of human normal and malignant glial cells. Biochem. J. 172: 443-456.

(40) Gospodarowicz, D; Greenberg, G.; Bialecki, H.; and Zelter, B. 1978. Factors involved in the modulation of cell proliferation in vivo and in vitro: The role of fibroblast and epidermal growth factors in the proliferative response of mammalian cells. In Vitro 14: 85-118.

(41) Gospodarowicz, D.; Ill, C.R.; and Birdwell, C.R. 1977.
      Effects of fibroblast and epidermal growth factors on
      ovarian cell proliferation in vitro. I. Characterization
      of the response of granulosa cells to FGF and EGF. Endo-
      crinology 100: 1108-1115.

(42) Gould, R.M., and Dawson, R.M.C. 1976. Incorporation of
      newly formed lecithin into peripheral nerve myelin. J.
      Cell Biol. 68: 480-496.

(43) Griffin, J.W., and Price, D.L. 1981. Segmental demyelination
      in experimental IDPN and hexacarbon neuropathies: Evidence for
      an axonal influence. Lab. Invest., in press.

(44) Grossman, Y.; Parnas, I.; and Spira, M. 1979. Ionic
      mechanisms involved in differential conduction of action
      potentials at high frequency in a branching axon. J.
      Physiol. 295: 307-322.

(45) Güldner, F.H., and Wolff, J.R. 1973. Neurono-glial synap-
      toid contacts in the median eminence of the rat: Ultra-
      structure, staining properties and distribution on tany-
      cytes. Brain Research 61: 217-234.

(46) Haley, J.E., and Ledeen, R.W. 1979. Incorporation of
      axonally transported substances into myelin lipids. J.
      Neurochem. 32: 735-742.

(47) Hall, S.M., and Gregson, N.A. 1971. The in vivo and
      ultrastructural effects of injection of lysophosphatidyl
      choline into myelinated peripheral nerve fibres of the
      adult mouse. J. Cell Science 9: 769-789.

(48) Hökfelt, T., and Ljungdahl, A. 1971. Light and electron
      microscopic autoradiography on spinal cord slices after
      incubation with labelled glycine. Brain Res. 23: 189-194.

(49) Itoyama, Y.; Sternberger, N.; Webster, H.; Quarles, R.;
      Cohen, S.; and Richardson, E. 1980. Immunocytological ob-
      servations on the distribution of myelin-associated glyco-
      protein and myelin basic protein in multiple sclerosis
      lesions. Ann. Neurol. 7: 167-177.

(50) Jacobs, J. 1977. Penetration of systemically injected horse-
      radish peroxidase into ganglia and nerves of the autonomic
      nervous system. J. Neurocytol. 6: 607-618.

(51) Kanazawa, I.; Iversen, L.L.; and Kelly, J.S. 1976. Gluta-
      mate decarboxylase activity in the rat posterior pituitary,
      pineal gland, dorsal root ganglion and superior cervical
      ganglion. J. Neurochem. 27: 1267-1269.

(52) Kehoe, J. 1976. Electrogenic effects of neutral amino
      acids on neurons of Aplysia Californica. Cold Spring
      Harbor Harbor Symp. Quant. Biol. 40: 145-155.

(53) Kelly, J.S., and Dick, F. 1976. Differential labelling of glial cells and GABA-inhibitory interneurons and nerve terminals following the microinjection of (^{3}H) β-alanine, (^{3}H) DABA and (^{3}H) GABA into single folia of the cerebellum. Cold Spring Harbor Symp. Quant. Biol. 49: 93-106.

(54) Kiang, W.-L.; Crockett, C.P.; Margolis, R.K.; and Margolis, R.U. 1978. Glycosaminoglycans and glycoproteins associated with microsomal subfractions of brain and liver. Biochem. 17: 3841-3848.

(55) Kimmelberg, H.K.; Narumi, S.; Biddlecome, S.; and Bourke, R.S. 1978. Na, K-ATPase, ^{86}Rb$^+$ transport and carbonic anhydrase in isolated brain cells and cultured astrocytes. In Dynamic Properties of Glial Cells, eds. E. Schoffeniels et al., pp. 347-357. New York: Pergamon Press.

(56) Kuffler, S.W.; Nicholls, J.G.; and Orkand, R.K. 1966. Physiological properties of glial cells in the central nervous system of amphibia. J. Neurophysiol. 29: 768-787.

(57) Kuffler, S.W., and Potter, D.D. 1964. Glia in the leech central nervous system; physiological properties and neuron-glia relationship. J. Neurophysiol. 27: 290-320.

(58) Kukes, G.; Elul, R.; and DeVellis, J. 1976. The ionic basis of the membrane potential in a rat glial cell line. Brain Res. 104: 71-92.

(59) Larrabee, M.G., and Klingman, J.D. 1962. Metabolism of glucose and oxygen in mammalian sympathetic ganglia at rest and in action. In Neurochemistry, eds. K.A.C. Elliott, I. Page, and J.H. Quastel, 2nd ed., pp. 150-176. Springfield, IL: C.C. Thomas.

(60) Lasek, R.J.; Gainer, H.; and Barker, J.L. 1977. Cell-to-cell transfer of glial proteins to the squid giant axon. J. Cell Biol. 74: 501-523.

(61) LeBeux, Y.J. 1972. An ultrastructural study of the neurosecretory cells of the medial vascular prechiasmatic gland. II. Nerve endings. Z. Zellforsch. 127: 439-461.

(62) Ledbetter, M., and Lubin, M. 1977. Control of protein synthesis in human fibroblasts by intracellular potassium. Exp. Cell Research 105: 223-236.

(63) Leonhardt, H., and Backhus-Roth, A. 1969. Synapse-like contacts between the intraventricular axonal end bulbs and the apical plasmalemma of ependyma (rabbit). Z. Zellforsch. 97: 369-376.

(64) Lindsay, R.M. 1979. Adult rat brain astrocytes support survival of both NGF-dependent and NGF-insensitive neurones. Nature 282: 80-82.

200   State of the Art Report; P.A. Walicke et al.

(65) Margolis, R.U., and Margolis, R.K. 1979. Complex Carbo-
      hydrates of Nervous Tissue. New York: Plenum Press.

(66) Martinez-Hernandes, A.; Bell, P.; and Norenberg, M.D.
      1977. Glutamine synthetase: glial localization in brain.
      Science 195: 1356-1358.

(67) Matus, A.; DePetris, S.; and Raff, M.C. 1973. Mobility
      of concanavalin-A receptors in myelin and synaptic mem-
      branes. Nature, New Biol. 244: 278-280.

(68) McDonald, T.; Sachs, H.; Orr, C.; and Ebert, J. 1972.
      External potassium and baby hamster kidney cells: Intra-
      cellular ions, growth, DNA synthesis and membrane potential.
      Develop. Biology 28: 290-303.

(69) Means, A., and Dedman, J. 1980. Calmodulin - an intracel-
      lular calcium receptor. Nature 285: 73-77.

(70) Mirsky, R.; Winter, J.; Abney, E.R.; Pruss, R.M.;
      Gavrilovic, J.; and Raff, M.C. 1980. Myelin-specific
      proteins and glycolipids in rat Schwann cells and oligo-
      dendrocytes in culture. J. Cell Biol. 84: 483-494.

(71) Morell, P., and Norton, W.T. 1980. Myelin. Sci. Am.
      242: 88-118.

(72) Moya, F.; Bunge, M.B.; and Bunge, R.P. 1980. Schwann
      cells proliferate but fail to differentiate in defined
      medium. Proc. Natl. Acad. Sci. (USA) 77: 6902-6906.

(73) Nicholson, C. 1980. Dynamics of the brain cell micro-
      environment. Neurosci. Res. Prog. Bull. 18: 177-322.

(74) Norton, W.T. 1977. Chemical pathology of diseases
      involving myelin. In Myelin, ed. P. Morell, pp. 383-413.
      New York: Plenum Press.

(75) Ochs, S.; Erdman, J.; Jersild, R.A.; and McAdoo, V.
      1978. Routing of transported materials in the dorsal
      root and nerve fiber branches of the dorsal root ganglion.
      J. Neurobiology 9: 465-481.

(76) Olsson, Y. 1975. Vascular permeability in the peripheral
      nervous system. In Peripheral Neuropathy, eds. P.J. Dyck,
      P.K. Thomas, and E. Lambert, pp. 190-200. Philadelphia:
      Saunders.

(77) Oppenheim, R.; Chu-Wang, I.; and Maderdrut, J. 1978.
      Cell death of motoneurons in the chick embryo spinal cord.
      III. The differentiation of motoneurons prior to their in-
      duced degeneration following limb-bud removal. J. Comp.
      Neurology 177: 87-112.

(78)   Patterson, P.H.   1978.   Environmental determination of
       autonomic neurotransmitter functions.   Ann. Rev. Neurosci.
       1: 1-18.

(79)   Raff, M.; Fields, K.; Hakomari, S-I.; Mirsky, R.; Pruss,
       R.; and Winter, J.   1979.   Cell-type-specific markers for
       distinguishing and studying neurons and the major classes
       of glial cells in culture.   Br. Res. 174: 283-308.

(80)   Raine, C.S.; Wisniewski, H.; Dowling, P.C.; and Cook, S.D.
       1971.   An ultrastructural study of experimental demyelina-
       tion and remyelination.   IV. Recurrent episodes and periph-
       eral nervous system plaque formation in experimental en-
       cephalomyelitis.   Lab Invest. 25: 28-34.

(81)   Sabri, M.I.; Bone, A.H.; and Davison, A.N.   1974.   Turn-
       over of myelin and other structural proteins in developing
       rat brain.   Biochem. J. 142: 499-507.

(82)   Salzer, J.L., and Bunge, R.P.   1980.   Studies of Schwann
       cell proliferation.   I. An analysis in tissue culture of
       proliferation during development, Wallerian degeneration,
       and direct injury.   J. Cell Biol. 84: 739-752.

(83)   Savage, C.R., and Cohen, S.   1973.   Proliferation of
       corneal epithelium induced by epidermal growth factor.
       Exp. Eye Res. 15: 361-366.

(84)   Schon, F.S., and Kelly, J.S.   1974.   Autoradiographic
       localization of (^{3}H) glutamate over satellite cells.
       Brain Res. 66: 275-288.

(85)   Schon, F.S., and Kelly, J.S.   1974.   The characterization
       of (^{3}H) GABA uptake into the satellite cells of rat sensory
       ganglia.   Brain Res. 66: 289-300.

(86)   Singer, M., and Salpeter, M.M.   1966.   The transport of
       (^{3}H)-L-histidine through the Schwann and myelin sheath
       into the axon including a reevaluation of myelin function.
       J. Morphol. 120: 281-316.

(87)   Suzuki, K.; Kamoshita, S.; Eto, Y.; Tourtellotte, W.W.;
       and Gonatas, J.O.   1973.   Myelin in multiple sclerosis.
       Arch. Neurol. 28: 293-297.

(88)   Thompson, R.H.S.   1973.   Fatty acid metabolism in multiple
       sclerosis.   Biochem. Soc. Symp. 35: 103-111.

(89)   Treherne, J.E.   1967.   Axonal function and ionic regula-
       tion in insect central nervous tissues.   In Insects &
       Physiology, eds. J.W.L. Beament and J.E. Treherne,
       pp. 175-188.   Edinburgh and London: Oliver & Boyd.

(90)   Treherne, J.E.   1974.   The environment and function of
       insect nerve cells.   In Insect Neurobiology, ed., J.E.
       Treherne, pp. 187-244.   Amsterdam: North Holland.

(91) Tucek, S.; Zelena, J.; Ge, J.; and Vyskocil, F. 1978.
      Choline acetyltransferase in transected nerves, denervated
      muscles and Schwann cells of the frog: Correlation of bio-
      chemical, electron microscopical and electrophysiological
      observations. Neuroscience 3: 709-724.

(92) Varon, S.; Raiborn, C.; and Burnham, P.A. 1974. Selec-
      tive potency of homologous ganglionic non-neuronal cells
      for the support of dissociated ganglionic neurons in
      culture. Neurobiology 4: 231-252.

(93) Varon, S.S., and Somjen, G.G. 1979. Neuron-glia inter-
      actions. Neurosci. Res. Progr. Bull. 17: 6-183.

(94) Villegas, J. 1978. Cholinergic properties of satellite
      cells in the peripheral nervous system. In Dynamic
      Properties of Glial Cells, eds. E. Schoffeniels et al.,
      pp. 207-215. New York: Pergamon Press.

(95) Vlodavsky , I., and Gospodarowicz, D. 1979. Structural
      and functional alterations in the surface of vascular
      endothelial cells associated with the formation of a con-
      fluent cell monolayer and with the withdrawal of a fibro-
      blast growth factor. J. Supramolecular Structure 12:
      73-114.

(96) Walsh, J.M., Bowery, N.G.; Brown, D.A.; and Clark, J.B.
      1974. Metabolism of γ-aminobutyric acid (GABA) by
      peripheral nervous tissue. J. Neurochem. 22: 1145-1147.

(97) Weiss, P.A., and Mayr, R. 1971. Organelles in neuro-
      plasmic (axonal) flow: Neurofilaments. Proc. Natl. Acad.
      Sci. (USA) 68: 846-850.

(98) Westermark, B., and Wasteson, Å. 1975. The response of
      cultured human normal glial cells to growth factors.
      Adv. Metab. Disorders 8: 85-100.

(99) White, C.J.B. 1979. Identification of glycosaminogly-
      cans in nerve terminals. J. Neurol. Sci. 41: 261-269.

(100) Wolff, J.R.; Rickmann, M.; and Chronwall, B. 1979. Axo-
      glial synapses and GABA-accumulating glial cells in the
      embryonic neocortex of the rat. Cell Tiss. Res. 201:
      239-248.

(101) Young, J.A.C.; Brown, D.A.; Kelly, J.S.; and Schon, F.S.
      1973. Autoradiographic localization of sites of (^{3}H) γ-
      aminobutyric acid accumulations in peripheral autonomic
      ganglia. Brain Res. 63: 479-486.

Neuronal-glial Cell Interrelationships, ed. T.A. Sears, pp. 203-216.
Dahlem Konferenzen 1982. Berlin, Heidelberg, New York: Springer-Verlag.

Mechanisms of Tissue Damage in the Nervous System

B. G. W. Arnason* and B. H. Waksman**
*Dept. of Neurology, University of Chicago, Chicago, IL 60637
**National MS Society, New York, NY 10017, USA

INTRODUCTION

This paper will emphasize immunologically mediated tissue
damage within nervous tissue. Certain conditions have been
arbitrarily selected as prototypes around which discussion of
mechanisms can be based.

SYDENHAM'S CHOREA

Sydenham's chorea (SC) is characterized by purposeless in-
voluntary movements, weakness, and emotional lability. The
movements disappear during sleep, a characteristic shared by
other movement disorders and may be damped down by rest, seda-
tion, and by glucocorticoids. Behavioral abnormalities, typically fits of crying, giddiness, or temper tantrums, are common in SC and may progress to psychosis. The electroencephalo-
gram (EEG) may exhibit diffusely abnormal slow wave activity
indicative of a widespread process within brain. Symptoms
endure for anywhere from a week up to two years.

No pathologic substratum for SC is known. SC is rarely fatal
and the handful of autopsied cases all had fulminating rheu-
matic fever (RF) so that no specificity can be ascribed to the
pathologic features (chiefly cerebral vasculitis) noted. There

is reason to believe that the caudate nucleus is a major site
of disease because this region is selectively damaged in other
choreic syndromes, e.g., Huntington's disease. SC may occur
in pure form or superimposed upon RF. In either case it is
tied to antecedent Group A streptococcal infection.

Cell membranes from Group A streptococci contain proteins which
cross-react immunologically with cardiac and skeletal muscle,
valvular tissue, kidney, and skin. Antibodies (Abs) which bind
to heart and skeletal muscle sarcolemmal membrane and are absorbable with streptococcal membrane are present in the sera
of acutely ill RF patients. Antibody (Ab) titers correlate
with clinical severity, fall with remission, and rise with re-
currence. These findings have led to the impression that autoimmunity contributes to the pathogenesis of RF although how
this occurs remains enigmatic. Mere presence of antiheart Abs
does not suffice; antiheart Abs are seen in other conditions
(e.g., myocardial damage) and do not, under these circumstances,
cause disease. Possibly vascular damage caused by streptococcal
toxins paves the way for a noxious action of antiheart Abs.
Alternatively cell-mediated immunity (CMI) directed against
heart could play a part. The presence of antiheart Abs in RF
prompted a search for antibrain Abs in SC.

Husby et al. (6) using immunofluorescence techniques found Abs
against caudate and subthalamic neurons (some sera also reacted
against hypoglossal and trigeminal nuclei and against cortex)
in 47 percent of serums from SC children. Titers correlated
with severity and duration of attacks. Absorption with caudate
neurons or streptococcal membranes removed the Ab. The Ab was
not absorbable with sarcolemma indicating nonidentity of antineuronal and antiheart Abs. The Ab was concentrated in medium
to large sized neurons, was bound to cytoplasmic constituents,
was chiefly of IgG_1 class, and reacted with complement but not
with the ligand binding sites of receptors for serotonin,
dopamine (DA), norepinephrine, acetylcholine (Ach), or gamma-
aminobutyric acid (GABA).

Circulating immune complexes (IC's) were found in virtually
all SC patients at levels 6 to 7 times background; only 60
percent of RF patients showed IC's. It was suggested that
IC's might disrupt the blood-brain barrier (BBB) thus facilitating passage of antineuronal Abs into brain and Ab interaction with neurons. The choroid plexus (CP), ordinarily im-
pervious to proteins, is a site of preferential deposition of
IC's as has been shown in systemic lupus erythematosus (SLE)
and in its animal model the NZB/NZW mouse. A damaged CP might
favor passage of Abs into the cerebrospinal fluid (CSF) between which and brain parenchyma no barriers exist.

These observations invite comment. In terms of disease pathogenesis, the Ab binds to an unlikely site (i.e., intracytoplasmic) in contrast to sarcolemmal Ab which reacts with membrane.
Methodologic problems could exist here since cytoplasmic staining is more readily detected in tissue sections than is membrane staining. Possibly membrane bound antigen (An) remains
to be detected. The situation in myasthenia gravis (MG) comes
to mind. It was long known that Ab to cytoplasmic skeletal
muscle elements could be detected in older myasthenics. The
fire behind this smoke came to light years later when Ab directed against the Ach receptor of the myoneural junction (AchR)
was shown to be responsible for MG.

A second point relates to the possibility that immunologically
mediated brain dysfunction may be triggered by cross-reactivity
between brain components and infectious agents. This possibility
merits exploration in affective disorders and in schizophrenia.

SC is reversible; it follows that immune mediated dysfunction
need not lead to neuronal death. Binding of Ab to membrane
components alters function in other diseases. In Graves Disease
thyroid hyperfunction results from the action of thyroid stimulating Ab which is directed against the thyroid stimulating
hormone (TSH) receptor and occludes TSH binding either because
the ligand binding site and the immunoglobulin binding site are

the same, because of steric hindrance, or by antigenic modulation. In chronic thyroiditis, a second entity affecting the
same organ characterized clinically by thyroid hypofunction,
Abs, are directed against thyroglobulin, microsomes, second
colloid antigen, and a surface component distinct from the TSH
receptor. While the role of these Abs in thyroiditis is incom-
pletely elucidated, an important point can be made. Abs directed
against different determinants on a single target can eventuate
in hyper- or hypofunction. Physostigmine, an Ach agonist,
diminishes chorea; it follows that an Ab which blocked AchR
could cause chorea. Conversely DA augments chorea and an Ab
which activated the DA receptor could cause chorea.

PARANEOPLASTIC SYNDROMES
A partial listing is given in Table 1. In certain of these
entities immune mechanisms clearly operate (e.g., MG). It is
reasonable to suspect that immune mechanisms operate in the
others since inflammatory responses have been reported in patients dying acutely with the paraneoplastic syndromes listed.
Other paraneoplastic cases have shown only neuronal destruction, but the process could have burned itself out by the time
death occurred. Note that most paraneoplastic syndromes also
occur in the absence of tumor. Oat cell tumors and neuroblas-
tomas are of neural crest origin which may favor antigenic
cross-reactivity between tumor Ans and neural elements.

TABLE 1 - Paraneoplastic Syndromes of muscle, nerve, and brain

Disease	Associated Tumors	No tumor
Myasthenia	Thymoma, breast	most
Polymyositis	breast, lung, ovary	most
Polyneuritis	lymphoma	most
Lambert-Eaton	oat cell	30%
Cerebellar degen.	oat cell	none
Sensory neuropathy	oat cell	5%
Dysautonomia	oat cell	80%
Encephalomyelitis	oat cell, lymphoma	?
Opsoclonus-myoclonus	neuroblastoma	50%

SYSTEMIC LUPUS ERYTHEMATOSUS

SLE is characterized by multiple immune abnormalities and inflammatory destructive processes in joints, skin, kidneys,
heart, lungs, and brain. Multiple autoAbs have been documented
in SLE; some are important in the pathogenesis of tissue damage.
Most series list a 50 to 60 percent incidence of neurologic
symptoms and/or signs in SLE. CNS disease in SLE accounts for
15 percent of deaths. Five year survival of patients with
CNS-SLE is only 50%. Behavioral disturbances are the most frequently encountered manifestations of CNS involvement. These
are varied - organic psychoses, affective disorders, and schizo-
phreniform states being the most common. Attacks of psychiatric
disorder seldom endure beyond 6 months. Recurrences are frequent and the clinical picture may vary between attacks. The
EEG shows diffuse slowing indicative of widespread disease.
After psychoses, seizures are the second most common CNS sign;
usually these are grand mal, often with localizing signs.
Hemorrhages also occur as may "migraine." Examination of SLE
brains has proven unrewarding. Small infarcts may be present.
These may account for the wide array of focal signs seen but
not for an acute psychosis nor in most instances for seizures.

IgG, IgM, and complement are deposited in the CP in SLE presumably as IC's. IC's in CP could alter CSF formation leading
to water, electrolyte, or pH alterations which could derange
cortical function. This notion may have merit but psychotic
states are common early in SLE when glomerular function is
normal. There is no correlation between CNS and renal involvement as would be expected if two sinks for IC's were present
concurrently. Damage to the CP-CSF barrier could facilitate
passage of Ab into the CSF. Were this occurring, total protein in CSF would increase; this is observed only erratically
in CNS-SLE. The impermeability of brain to immunoglobulin (Ig)
should be placed in perspective. In rabbits, immunized system-
atically with protein An, intracisternal injection of An induces
widespread Arthus lesions. Some Ig (and complement) enters
brain normally; both are found in normal CSF in minute quantities. An immune reaction even if miniscule initially will

disrupt the BBB and permit a cascading reaction to go forward.
It is not necessary therefore for the BBB to be altered for an
An-Ab reaction to proceed in brain. A possible exception could
relate to Igs of IgM class.

Abs to the formed elements of blood are characteristic of SLE.
Might anti-brain Abs also be present? Quismorio and Friou (9)
found Abs to neuronal cytoplasm in 41% of CNS-SLE cases (com-
pare with SC). The Abs reacted with other tissues but were much
less commonly detected in SLE without CNS involvement or in
controls. Bluestein and Zvaifler (3) found that lymphocytotoxic
Ab (common in SLE) cross-reacted with brain. There was some
correlation of lymphocytotoxic Ab titer and the presence of CNS-
SLE. Similarly in NZ mice, IgM class Abs cytotoxic to disasso-
ciated cerebellar cells are found sporadically. The Abs cross-
react with kidney but not with T-cells. The possibility that
antibrain Abs may have a role in CNS-SLE merits study. Mention
should be made of early reports that Abs reactive with neuronal
nuclei are present in schizophrenia.

In SLE immune regulation is deranged; suppressor lymphocyte
(S-cell) function is defective. In SLE certain histocompat-
ability (HLA) Ans are overrepresented. These provide markers
for immune response (Ir) genes which predispose to SLE. C type
viruses may be implicated in murine SLE; in canine SLE a role
for virus as an inducer has been proposed. Fragmentary evidence
along the same lines exists in human SLE. An extrinsic agent
in a favorable terrain (in an immunogenetic sense) may trigger
an alteration in immune regulation which permits expression of
an array of autoAbs.

In polyarteritis nodosa (PN), peripheral nerve involvement is
common. Most often this presents as a mononeuritis multiplex
but ascending polyneuropathy is also observed. Lesions consist
of inflammatory occlusions of small and medium vessels. CNS
involvement (in PN) may present as headaches, seizures, focal
lesions, or as hemorrhage. Frank psychoses, common in SLE, are

rare. This argues that vasculitis is not responsible for the
mental derangements observed in SLE.

MYASTHENIA GRAVIS

MG is an autoimmune disease (1). Circulating Ab and CMI directed against AchR are found and are disease specific. MG
is associated with certain HLA-Ans (B8 and DW3 in young patients; A2 in elderly patients and those with thymomas), indicating that Ir genes have a role in the disease. What generates the autoimmune response is not known; viral infection has
been proposed but no data exist. A subset of lymphocytes which
may express a nicotinic AchR is depleted from the blood in MG -
again suggesting cross-reactivity between lymphocytic and tissue antigens.

An experimental model for MG has been developed. This has
been best studied in rats immunized with AchR. Disease occurs
in two waves. First wave disease is characterized by low Ab
titer and by stripping of the subsynaptic apparatus by macrophages. After this burst of activity, the inflammatory response subsides and the animal recovers. Ab titers continue
to rise and after 3 to 4 weeks second wave disease appears.
Ab and complement bind to AchR and the complex is taken into
the muscle by pinocytosis stripping the membrane of AchR.
There is no inflammatory response, i.e., receptor modulation
occurs without inflammation. When serum from animals with
second wave disease is given to virgin recipients, first wave
disease develops. This observation remains unexplained. MG
provides a model for other diseases in which membrane receptors for transmitters, hormones, growth factors, viruses, etc.,
might be stripped. In rare cases of insulin resistant diabetes,
Abs to the insulin receptor impair insulin binding. Interestingly Ab binding initially provides the full range of insulin
actions (i.e., these Abs resemble anti-TSH-Ab in Graves disease),
but after some hours postreceptor adaptation occurs which renders the cells refractory to insulin. Entities involving the
CNS in which similar mechanisms operate may await discovery.

EXPERIMENTAL ALLERGIC ENCEPHALOMYELITIS (EAE) AS A MODEL FOR
MULTIPLE SCLEROSIS (MS)
Several animal models suggestive of MS have been developed.
Included are selected viral infections, EAE, and combinations
of the two. In some virus models, notably those involving
distemper and Theiler's disease, chronic lesions may be due
to EAE triggered by viral infection.

Pathologic changes in acute EAE consist of perivascular cuffing
and meningitis throughout the CNS particularly in periventricular white matter and subpially in the spinal cord. The cuffs
contain small and large lymphocytes, some plasma cells, and a
preponderance of monocytes, which invade the adjacent parenchyma
and transform to mature macrophages. Myelin breakdown of two
types is seen in regions of cell infiltration; swelling and
vesicular breakdown, and stripping of myelin with phagocytosis
by macrophages.   Inflammatory cell influx precedes clinical
signs by 1-3 days. Hyperacute lesions show superimposed vascular necrosis with leakage of plasma constituents, hemorrhage,
and neutrophil infiltration.

Chronic relapsing EAE can be induced in guinea pigs immunized
with myelin at weaning (when regulatory mechanisms are in transition from juvenile to adult forms) and in mouse strains such
as SJL, which exhibit abnormal immune regulation (8,10). In
chronic EAE, lesions of different ages are present simultaneously. In addition, there may be continuing demyelination at
the edges of old lesions. Large confluent plaques are seen in
which axons are spared and there is fibrous gliosis. Remyelination is observed, both by oligodendroglia (OG) and, near
root entry zones in the cord, by Schwann cells. The lesions
resemble those of MS. EAE lesions produce a local breakdown
of the BBB and EAE is greatly exaggerated in regions where the
BBB is artificially reduced, e.g., by local trauma.

EAE can be transferred to normal recipients by T-cells from
sensitized donors. Transfer is enhanced by exposing the cells

in vitro to mitogen or to myelin antigen before transfer. Circulating lymphocytes, in immunized animals, show an increase
in "early" or "avid" T-lymphocytes; these fall sharply with
onset of disease. Cells washed from the meninges and brain
in which acute lesions are present show an increase in these
cells. In chronic EAE, a decrease in circulating early T-cells
is associated with each relapse. The major histocompatibility
locus related genetic restriction of EAE production among in-
bred strains may reflect restrictions of antigen recognition.

EAE may be prevented by immunosuppressive and antiinflammatory
agents (5). As with other CMI reactions, EAE is inhibited by
diets high in polyunsaturated fats and by other agents which
raise levels of prostaglandins of the E series. EAE may also
be suppressed by injections of myelin or myelin antigens. Myelin basic protein (MBP) and synthetic MBP analogues have been
used in recent studies of this type. The effect may depend on
generation of An-specific S-cells. Spontaneous termination of
an EAE attack is also associated with generation of S-cells.

Major problems surround the EAE model. The first concerns the
antigen. MBP induces acute EAE in several species and elicits
CMI type skin reactions. Since MBP is well characterized, it
has been used extensively in functional and other studies for
over 20 years. Effective MBP sequences differ between species,
and there may be multiple encephalitigenic determinants in MBP
for a single species. Determinants effective in suppression
may differ from those responsible for CMI and disease.

MBP may not be the only or even the principal antigen involved
in chronic EAE. Repeated injections of myelin glycolipids are
reputed to produce autoimmune CNS or PNS lesions, and immunization with myelin proteolipid protein has been claimed to cause
EAE. The possibility that two distinct determinants (possibly
two distinct Ans) may be required to induce separate simultaneous cell-mediated and antibody responses is currently being
investigated. MBP alone does not induce chronic EAE in guinea
pigs or mice; yet MBP suppresses both the initial attack and

relapses in guinea pigs sensitized with myelin. This supports
the notion that two antigens may be involved. Emphasis on
myelin should not obscure the possibility that OG cell antigens
may serve to elicit a CMI reaction or may serve as target in
an Ab mediated event leading to demyelination.

A second problem concerns the events leading to demyelination.
There seems little doubt that the cellular inflammation represents a T-cell-mediated event with a secondary infiltration of
macrophages which destroy myelin. The concept that activated
macrophages entering parenchyma destroy myelin has been chal-
lenged by the observation that there is little demyelination
in EAE of guinea pigs sensitized with MBP and a great deal in
EAE of animals sensitized with whole myelin, although macro-
phage invasion appears comparable in the two types of EAE.
Obvious questions are: a) Are there additional elements in
the vascular lesion associated with the initial T-cell reaction?
Protease inhibitors inhibit lesion formation while aprotinin
enhances it. Here one thinks of possible involvement of complement, plasmin, or clotting systems in the vascular process.
Indeed inhibitors of coagulation inhibit EAE. b) Are additional
elements involved in destruction of OGS and myelin? In the
eyes of rabbits injected intravitreally with lymphokine containing supernatants from activated lymphocytes, mononuclear inflammation appears around vessels but little myelin breakdown
occurs. If Ab to myelin constituents is injected simultaneously,
demyelination may be extensive. Ab alone does not demyelinate.
It is hard to escape the conclusion that an Ab-dependent cellular cytotoxicity (ADCC) event is superimposed on the inflammation resulting from T-cell mediated immunity when whole myelin serves as the immunizing agent. The Ans responsible for
the two events may differ, e.g., MBP for CMI and galactocerebroside for Ab. This view is supported by experiments in which a
typical CMI lesion is produced in dorsal roots of tuberculin
sensitized rats injected locally with PPD, but there is demyelination only if the animals also possess Ab against galactocerebroside (7). Details of the ADCC are still unknown.

Emphasis has been placed on release of proteolytic enzymes by participating macrophages. As noted, protease inhibitors inhibit EAE, but it is unclear whether inhibition is exerted on myelin breakdown or earlier events at the level of the vessel wall. c) A third problem concerns the role in chronic EAE of the immunoregulatory system and the defect in juvenile animals or in certain strains of mice that permits a chronic, oscillating expression of CMI to myelin antigen. Equally important is the nature of "trigger" events that boost or renew this expression.

Many problems raised in the study of animal models may be resolvable by study of systems consisting of single cell populations or combinations of two or more populations. Such systems permit analysis of cell signalling for proliferation, myelination, gliosis, etc. They can be used to analyze the effector arm of the EAE model, i.e., the cytotoxic effects of T-cells, activated macrophages, and the mediators and enzymes produced by each, during different types of myelin breakdown. The effects of ADCC with Abs differing in molecular class and in specificity can be studied, as well as the role of non-Ig myelinotoxic factors. Employment of monoclonal antibodies generated by the hybridoma technique should facilitate this task. The nature of conduction defects with different degrees of immunologically mediated demyelination or in the presence of neuroelectric blocking factors (Ab or others?) can be determined. For a detailed discussion of EAE the recent monograph edited by Davison and Cuzner should be consulted (5).

MULTIPLE SCLEROSIS

MS is an inflammatory disease; the lesions resemble those of chronic EAE. Lesions are perivenular initially; larger ones form by coalescence of smaller ones plus activity at plaque margins. OGs disappear from the MS plaque; it is unclear whether comparable destruction of OGs occurs in EAE. Remyelination is more evident in EAE than in MS. HLA-A3, B7, and DW2 are increased in MS pointing to an Ir gene effect. Epidemiologic evidence suggests a role for an extrinsic factor,

possibly viral, though whether the disease is a "slow virus
disease" or an autoimmunity (possibly triggered by a virus or
bacterium) remains unknown. Characteristic of MS is the presence of oligoclonal IgG_1 bands in the CSF. Similar bands are
present in subacute sclerosing panencephalitis; in this disease they have been shown to be anti-measles Ab. All attempts
to define an An, viral or neural, against which the MS-IgG
bands are directed have failed.

A search has been made for anti-MBP Abs in MS and for the
presence of CMI directed against this same An. The issue is
unresolved; positive and negative results have been reported.
In our view MBP is an unlikely antigen in MS although it may
well be in post-infectious encephalomyelitis. MBP injected
into MS patients leads to development of CMI to this material.
Response begins after 10 days to 3 weeks, i.e., the time which
would be expected for a primary take. This suggests that CMI
to MBP was not present prior to MBP injection. What then might
the antigen be? Consideration should be given to a membrane
receptor on OGs. S-cells disappear from the blood and CSF
during attacks of MS. Possibly this permits a sensitivity held
in check between attacks to declare itself. Alternatively the
"MS antigen" could be shared by S-cells and Ogs. Investigation
of these possibilities could prove fruitful. For detailed consideration of the immunology of MS the interested reader may
consult the recent monographs edited by Bauer, Poser, and Ritter
(2) and by Boese (4).

REFERENCES

(1) Arnason, B.G.W. 1980. Myasthenia Gravis. In Clinical
     Immunology, ed. C. Parker, vol. 2, pp. 1088-1105. Phila-
     delphia: Saunders.

(2) Bauer, H.J.; Poser, S.; and Ritter, G., eds. 1980. Pro-
     gress in Multiple Sclerosis Research. Berlin: Springer-
     Verlag.

(3) Bluestein, J.G., and Zvaifler, N.J. 1976. Brain reac-
     tive lymphocytotoxic antibodies in the serum of patients
     with systemic lupus erythematosus. J. Clin. Invest. 57:
     509-516.

(4) Boese, A., ed. 1980. Search for the Cause of Multiple
     Sclerosis and Other Chronic Diseases of the Central Ner-
     vous System. First International Hertie Foundation Sym-
     posium. Weinheim: Verlag Chemie.

(5) Davison, A.J., and Cuzner, M.L., eds. 1980. The Suppres-
     sion of Experimental Allergic Encephalomyelitis and Mul-
     tiple Sclerosis. London: Academic Press.

(6) Husby, G.; Williams, R.C., Jr.; Bersin, R.M.; and Lewis,
     M.K. 1979. Antineuronal antibodies in diseases affecting
     the basal ganglia particularly Sydenham's and Huntington's
     Chorea. In Clinical Neuroimmunology, ed. F.C. Rose, pp.
     90-105. Oxford: Blackwell.

(7) Lampert, P.W. 1980. Structural manifestations of immune
     phenomena resulting in demyelination. J. Neuropathol.
     Exper. Neurol. 39: 369 (Abstract).

(8) Lessman, H., and Wisniewski, H.M. 1979. Chronic relapsing
     experimental allergic encephalomyelitis. Clinical and
     pathological comparison with multiple sclerosis. Arch.
     Neurol. 36: 490-497.

(9) Quismorio, F.P., and Friou, G.J. 1972. Antibodies reac-
     tive with neurons in SLE patients with neuropsychiatric
     manifestations. Int. Arch. Allergy. 43: 740.

(10) Raine, C.S., and Stone, S.H. 1977. Animal model for
     multiple sclerosis. Chronic EAE in inbred guinea pigs.
     N.Y. State J. Med. 77: 1693-1696.

Neuronal-glial Cell Interrelationships, ed. T.A. Sears, pp. 217-228.
Dahlem Konferenzen 1982. Berlin, Heidelberg, New York: Springer-Verlag.

Mechanism of Viral Injury to the Nervous System

B. N. Fields* and H. L. Weiner**
*Dept. of Microbiology and Molecular Genetics,
 Harvard Medical School, Boston, MA 02115
**Dept. of Neuroscience G-408, Children's Hospital Medical Center,
 Boston, MA 02115, USA

Abstract. The mechanism of viral injury to the nervous system
is dependent on the interaction of both viral and host factors.
The initial factors involved include entry of virus, spread to
the central nervous system, and cell tropism within the CNS.
Specific patterns of viral injury include destruction of CNS
elements and/or alteration of their function, establishment of
persistent infections, and virally directed immune damage to
the nervous system. The molecular basis for some of these processes are now beginning to be elucidated.

INTRODUCTION

Viruses are capable of infecting the nervous system in a number
of distinct ways. Injury can occur rapidly with marked hemorrhage and with necrosis of neurons as seen in herpes simplex
encephalitis (3), or pathology of ganglia cells can occur without the production of overt infection as is seen in latent
varicella or herpes simplex infection of sensory or trigeminal
ganglia (26). Between these two extremes are numerous instances
of milder viral infections involving different cell types (neural, ependymal, glial) occurring as acute infections or, alternatively, as persistent infections. Standing by themselves are the
spongiform encephalopathies ("slow virus") in which degenerative
disease occurs due to transmissable agents that have never been
unequivocally isolated or visualized and that cause chronic

infections without detectable febrile reactions of host immune
response (8). In the present overview we will review some of
the general strategies used by different viruses to produce central nervous system injury, emphasizing those systems where some
insights into mechanisms exist, either viral or host. This review is thus clearly somewhat selective, stressing systems where
basic data provide a framework for discussion of mechanism.

ENTRY INTO THE HOST

Viruses enter the mammalian host primarily via the gastrointestinal tract (e.g., polio), the respiratory tract (e.g., measles),
direct inoculation by arthropods into the blood stream (equine
encephalitis), or by other breaks in the skin (rabies) (5).
Following introduction into these primary sites, viral replication takes place in primary target cells and, following amplifi-
cation, thus begins the process of spreading to more distal targets. Very little is known concerning molecular mechanisms
during the initiation of viral infection. It is probable, however, that the interaction of viruses with specific receptors on
surface epithelial cells or endovascular cells (see discussion
below) plays a central role at this phase in determining the
localization of viral infection (12). In addition, the nature
of the body fluids in the anatomic location of viral entry, e.g.,
respiratory secretion, gastrointestinal fluids, blood serum, determines in part whether a productive infection will be established (see Table 1). Recent studies with the reoviruses, for
example, have indicated that one of the surface polypeptides
(the µ1c polypeptide) is the target of proteolytic enzymes (25).
Loss in infectivity in some isolates results from digestion of
the µ1c polypeptide by gastrointestinal proteases. Resistance to
proteases in vivo thus allows primary infection to occur and
ultimately determines whether successful initiation of systemic
infection will occur. It is possible, but still somewhat controversial, that the influenza neuraminidase similarly plays a primary role at this phase of infection, allowing virus to reach the
primary site of viral multiplication successfully (4). Although
analogous specific proteins have not been recognized for other

Table 1 - Factors affecting entry through mucous membranes

Respiratory tract

IgA
Mucous coating-glycoprotein inhibitors
ciliary actions
alveolar macrophages

Gastrointestinal tract

IgA
acid
bile
proteases

Skin

anatomic barrier

---

viruses, it is likely that most, if not all, viruses have surface proteins that play similar primary roles in entry and
spread. Thus, the nature of these body fluids at surfaces as
well as throughout the host, represents a specific variable
involved in the mechanism of interaction with an infecting viral
pathogen.

How Do the Viral Agents Spread from the Site of $1^{\underline{o}}$ Multiplication
into the CNS?
An infecting virus that has successfully withstood the hostile
environment at the portal of entry must traverse body fluids
as well as cellular and membranous barriers to reach the CNS
(Fig. 1). The three major pathways of spread resulting in entry into the nervous system are: (a) bloodstream → blood brain
barrier → CNS; (b) peripheral nerves → retrograde spread → CNS;
(c) olfactory nerves → CNS (Table 2). Although the pathways of
spread by these three routes have been well described, especially
using the detection of viral antigen by fluorescent antibody
techniques, molecular mechanisms involved in these stages of
transport have not been proposed.

The nature of the dramatically different strategies chosen by
those viruses utilizing viremia (such as picornaviruses, reoviruses)

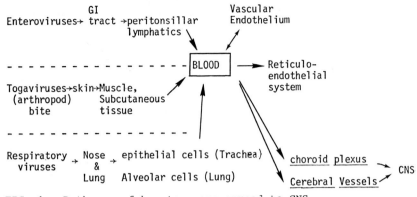

FIG. 1 - Pathways of hematogenous spread to CNS.

Table 2 - Pathways to CNS in natural infections

Route	Virus
A. Peripheral nerves	rabies
	herpes simirae
	herpes simplex
	varicella-zoster
B. Olifactory (uncertain)	togaviruses (?)
	rabies-bats (?)
	herpes simplex (?)
C. Blood	poliovirus and picornavirus

or retrograde neural spread (rabies, ? HSV) suggests possible
differences in mechanism. For example, viruses transported from
sites of primary multiplication into the blood could move freely
in plasma (polioviruses) or attached (via surface receptors,
either specific or non-specific) to cellular elements. Thus,
a virus that uses the hematogenous route must resist serum factors (complement, salts, etc.) and be transported either in the
fluids or cellular components of the blood. The specific mechanisms of entry into the host, spread from the primary site to
the CNS and ultimate entry beyond the anatomic barriers protect-
ing the brain from the blood have not been identified.

Once a Virus Has Entered into the Central Nervous System, What
Determines Spread and Ultimate Localization in Particular Cells?
Spread. The exact pathways of spread in the CNS are not well
understood, and the precise viral components and host components
responsible have not been identified for any virus. Most viruses
that reach the CNS via the choroid plexus, the olfactory mucosa,
or via intracerebral inoculation, are introduced into the sub-
arachnoid space and spread widely throughout contiguous spaces
(18). Different viruses have been shown to spread in a contiguous
fashion from meningeal or ependymal surfaces, in channels between
cells and processes, via glia, via cytoplasmic extensions of
neurons, or by mobile leukocytes. The specific host components
required by viruses during transit have not been identified.
For example, it is not known whether specific associations between viruses and cytoskeletal components (microtubules, microfilaments, intermediate filaments) might mediate viral transport. Herpes simplex, adenovirus, and reovirus types 1, 2, and
3 have all been found to associate with microtubules (2).

Cell tropism. The ultimate pattern of disease within the CNS
is determined to a considerable degree by the localization of
viruses within particular cells or regions of the brain. Certain viruses are highly localized in certain cell types (rabies
in neurons, reovirus type 1 in ependyma), while others appear
to be more pantropic (herpes simplex, measles in SSPE). Recent
studies using the model of the mammalian reoviruses have indicated
that the component of the virus responsible for the selective
localization is the viral hemagglutinin interacting in a highly
specific manner with cell surface receptors (32,33). Although
it has not been possible to prove that the hemagglutinin functions in an analogous way with other viruses, it appears quite
likely that this is the case. Thus the function of "binding"
to red blood cells (hemagglutination), in actuality, is detecting
the binding protein on the surface of the virus that is specialized to perform this function in vivo.

Another example of how the viral hemagglutinin (as well as other
viral surface glycoproteins) can affect the capacity of viruses
to produce disease has been shown by studies with myxoviruses
and paramyxoviruses. The hemagglutinin and other surface glycoproteins are responsible for adsorption to cells. With both
viruses, the two surface proteins exist as precursor molecules
that are cleaved by host proteases to active forms (15). Differences in susceptibility of the glycoproteins to proteases
have been shown to be important in host range and pathogenicity.
Avirulent strains have limited host range and most cell systems
are nonpermissive, i.e., they produce noninfectious virus containing uncleaved glycoproteins. With virulent strains, the
proteases of most host cells cause cleavage of the glycoproteins
and allow the virus to undergo multiple cycles of replication.
Thus, the combination of the viral glycoproteins with specific
host proteases helps determine the outcome of the infection.
Host factors are also important in determining which cells the
viruses localize. These host factors include genetic factors
that affect the capacity of cells of certain species to allow
permissive viral growth (24). In addition, age is a well-known
determinant of viral growth (10). In neither of these instances
is there a mechanism known that explains these striking effects.

It should be pointed out that it is quite probable that receptor
sites for viruses on the cell surface do not function solely as
receptors for a particular virus. The probability of a cell
generously providing receptors for viruses which inflict damage
and death is unlikely. The existence of cell surface binding
sites that serve both as phage or colicin receptors and as components of the uptake systems for low molecular weight compounds
(Vitamin B 12) and maltose or iron has been demonstrated in
several bacterial systems. Thus phages ϕ80, ϕ80h, T5, and colicin M all appear to bind to a specific surface receptor for the
siderophore, ferrichrome, and mutants lacking the receptor (tonA$^-$)
are both phage-resistant and unable to utilize ferrichrome (17).

It is known that receptor sites for a variety of viruses exist
on human lymphocytes (9), and presumably they are binding to a
physiologic receptor on the surface of the lymphocyte though
none have yet been identified. In terms of multiple sclerosis,
it has now been established that there are dramatic alterations of
peripheral blood T cell phenotypes during attacks (1,23). If
there were shared receptors on the lymphocyte and in the central
nervous system that were recognized by a particular virus, this
could serve as an initiating focus that would direct an autoimmune reaction against both peripheral blood lymphocytes and
the CNS. In this regard, recent work with reovirus has shown
that there is a "receptor" both on CNS tissue (neurons) and
peripheral blood lymphocytes specific for the hemagglutinin of
reovirus type 3 (31,33). The interaction of the virus via the
hemagglutinin results in a fatal encephalitis in the CNS (33).
It also alters immune function via the generation of suppressor
cells (7).

How Do Viruses Injure Cells?
The outcome of a viral infection of neural or virtually any cell
can vary from a highly cytocidal and destructive one to one with
a virtual absence of a detectable effect (28). Viral proteins
have been shown to shut off host cell protein synthesis rapidly
(polio, herpes). In other instances, this effect on host cell
metabolism may occur after a delayed period and not be as efficient (paramyxoviruses). With certain viruses (rabies) no
effects may be detected, while in LCM vital cell functions were
not affected but "luxury functions" were reduced (21). Viral
proteins have been found that cause rounding up of cells in
culture and have been called "toxins" (e.g., adenovirus fiber
proteins). In all instances, it is clear that there remains
little fundamental insight into the mechanism of injury. Recently, data have been generated to indicate more precise insights into the cellular target responsible for some of these
effects. For example, it has recently been shown that poliovirus infection mediates its effect on protein synthesis through
modification of the "cap" binding protein (29). In spite of

such isolated examples, the viral components responsible for
cell injury, how they mediate injury (on the surface of the
cells or in a particular intracellular compartment) are poorly
understood.

In addition to direct viral injury, it has been demonstrated in
several animal models that viral initiated, immune dependent
damage to the central nervous system can occur. Several models
include infections by arenaviruses of which lymphocytic chorio-
meningitis is an example. Immunosuppression of T cell responses
in adults or infected neonates prevents the development of CNS
disease. In addition, disease is transferable by immune T
lymphocytes but not by immune serum. The presumed mechanism is
destruction of infected CNS tissue by immune lymphocytes (20).
In another model, murine encephalomyelitis virus (Theiler virus)
is associated with an initial viral-induced flaccid paralysis
secondary to gray matter involvement. In animals that recover,
there is persistence of virus and then primary demyelination in
the spinal cord. The latter effects can be prevented by immunosuppression (16). Well described clinical examples of viral
associated immune mediated damage include post-infectious or
post-vaccinial encephalomyelitis and "virally triggered" damage
to peripheral nerve (Guillain-Barre). The mechanisms of injury
in these instances are unknown. It is possible that shared antigenic structures between virus and portions of the central nervous system, or alternatively, imbalances of immunoregulation
caused by the virus, could be related to these immunological events.

Why Do Some Viruses Persist in the Nervous System?
Numerous theories have been advanced to explain the capacity of
a number of different viral infections to persist in the nervous
system often in the face of a highly effective immune response.
In the case of DNA viruses and RNA viruses that contain reverse
transcriptase, DNA copies of the viral genome are produced that
can be integrated into the host cells as "cellular" genes (retro-
viruses) or that can replicate as extrachromosomal episomes
(herpes viruses). Thus, such viruses can exist in incomplete or

latent forms and can periodically reactivate complete virus
when the host immune response fails. Herpes simplex and varicella virus reside in sensory ganglia presumably as such incomplete virions. The retrovirus visna has been recovered from
choroid plexus of infected animals and undergoes sequential
changes during the persistent infection, resulting in antigenic
mutants that are no longer neutralized by the animals sera (19).

Other mechanisms invoked to explain the persistence of normally
lytic RNA viruses include viral mutation, defective interfering
viruses (DI's), and resistance to interferon (27). Ts mutants
and other viral mutants have been recovered from numerous persistently infected cell lines in vitro (22) and have been shown
to alter disease in vivo (6). Recent data with the reoviruses
suggests that a specific mutation in the viral gene product responsible for inhibition of protein synthesis (S4 gene) is the
critical mutation responsible for persistence (Ahmed and Fields,
unpublished). Experiments with ts mutants of reovirus in vivo
indicate that mutants can attenuate an acute efficient lytic
infection and allow illness to develop in a more progressive
fashion. The possibility that such mutants may exist in SSPE
is suggested by recent data indicating that there is an altered
host response to the M protein (11,30).

Defective interfering particles have also been suggested as
playing a possible role in persistent infection (13). In vivo
experiments in mice suggested that acute illness may be modified
by DI's but do not clearly establish a role in persistence.

Perspective
Over the last decade, revolutionary advances have occurred in
our understanding of both the genetics and molecular biology of
animal viruses as well as the mechanisms of host immunity (particularly in terms of the development of hybridoma technology
and its application as well as a more sophisticated understanding
of immune regulatory networks). The application of these advances for the study of such complex problems as the pathogenesis

of viral injury to the nervous system will provide the tools
needed to isolate the individual factors, both at viral and
host levels, responsible for the understanding of viral mechanisms of CNS disease.

REFERENCES

(1) Antel, J.P.; Arnason, B.G.W.; and Medof, M.E. 1978. Sup-
     pressor cell function in multiple sclerosis: correlation
     with clinical disease activity. Ann. Neurol. 5: 338-342.

(2) Babiss, L.E.; Luftig, R.B.; Weatherbee, J.A.; Weihing, R.;
     Ray, V.R.; and Fields, B.N. 1979. Reovirus serotypes dif-
     fer in their in vitro association with microtubules. J.
     Virol. 30: 863-874.

(3) Baringer, J.R. 1975. Herpes simplex virus infection of
     nervous tissue in animals and man. Progr. med. Virol. 20:
     1-26.

(4) Bucher, D., and Palese, P. 1975. The biologically active
     proteins of influenza virus: Neuraminidase. In The Influ-
     enza Viruses and Influenza, ed. E.D. Kilbourne, pp. 83-123.
     New York: Academic Press.

(5) Fenner, F.; McAuslan, B.R.; Mims, C.A.; Sambrook, J.; and
     White, D.O. 1974. The Biology of Animal Viruses, pp. 372-
     393. New York: Academic Press.

(6) Fields, B.N. 1972. Genetic manipulation of reovirus type
     3: A model for alteration of disease? New Engl. J. Med.
     287: 1026-1033.

(7) Fontana, A., and Weiner, H.L. 1980. Interaction of reo-
     virus with cell surface receptors. II. Generation of sup-
     pressor T cells by the hemagglutinin of reovirus type 3.
     J. Immunol. 125: 2660-2664.

(8) Gajdusek, D.C. 1977. Unconventional viruses and the ori-
     gin and disappearance of Kuru. Science 197: 943-960.

(9) Greaves, M.F. 1976. Virus "receptors" on lymphocytes.
     Scand. J. Immunol. 5: 8-23.

(10) Griffin, D.E.; Mullinix, J.; Narayan, O.; and Johnson, R.T.
     1974. Age dependence of viral expression: comparative patho-
     genesis of two rodent-adapted strains of measles virus in
     mice. Infect. Immun. 9: 690-695.

(11) Hall, W.W.; Lamb, R.A.; and Choppin, P.W. 1979. Measles
     and subacute sclerosing panencephalitis virus proteins:
     lack of antibodies to M protein in patients with subacute
     sclerosing panencephalitis. Proc. Natl. Acad. Sci. (USA)
     76: 2047-2051.

(12) Holland, J.J., and McLaren, L.C. 1961. The location and
     nature of enterovirus receptors in susceptible cells. J.
     Exp. Med. 114: 161-171.

(13) Huang, A.S., and Baltimore, D. 1970. Defective viral par-
     ticles and viral disease processes. Nature 226: 325-327.

(14) Johnson, R.T., and Mims, C.A. 1968. Pathogenesis of viral
     infections of the nervous system. New Eng. J. Med. 278:
     23-30, 84-92.

(15) Klenk, H.D. 1980. Viral glycoproteins: Initiators of in-
     fection and determinants of pathogenicity. In The Molecular
     Basis of Microbial Pathogenicity, eds. H. Smith, J.J. Skehel,
     and M.J. Turner, pp. 55-66. Weinheim: Verlag Chemie.

(16) Lipton, H.L., and Dal Canto, M.C. 1976. Theiler's virus-
     induced demyelination: prevention by immunosuppression.
     Science 192: 62-64.

(17) Meager, A., and Hughes, R.C. 1977. Virus Receptors. In
     Receptors and Recognition, eds. P. Cuatrecasas and M.F.
     Greaves, p. 152. London: Chapman and Hall.

(18) Mims, C.A. 1960. Intracerebral injections and the growth
     of viruses in the mouse brain. Brit. J. exp. Path. 41: 52-59.

(19) Narayan, O.; Griffin, D.E.; and Chase, J. 1977. Antigenic
     shift of visna virus in persistently infected sheep. Science
     197: 376-378.

(20) Nathanson, N.; Monjan, A.A.; Panitch, H.S.; Johnson, E.D.;
     Petursson, G.; and Cole, G.A. 1975. Virus-induced cell-
     mediated immunopathological disease. In Viral Immunology
     and Immunopathology, ed. A. Notkins, pp. 357-391. New
     York: Academic Press.

(21) Oldstone, M.B.A.; Perrin, L.H.; and Welsh, R.M. 1976
     Potential pathogenic mechanisms of injury in amyotrophic
     lateral sclerosis. In Amyotrophic Lateral Sclerosis, eds.
     J.M. Andrews, R.T. Johnson, and M.A.B. Brazier, pp. 251-
     262. New York: Academic Press.

(22) Preble, O.T., and Younger, J.S. 1975. Temperature-sensi-
     tive viruses and the etiology of chronic and inapparent in
     fections. J. Inf. Dis. 131: 467-473.

(23) Reinherz, E.L.; Weiner, H.L.; Hauser, S.L.; Cohen, J.A.;
     Distaso, J.A.; and Schlossman, S.F. 1980. Loss of sup-
     pressor T cells in active multiple sclerosis: Analysis with
     monoclonal antibodies. New Eng. J. Med. 303: 125-129.

(24) Rosenstreich, D.L.; O'Brien, A.D.; Groves, M.G.; and Taylor, B.A. 1980. Genetic control of natural resistance to infection in mice. In The Molecular Basis of Microbial Pathogenicity, eds. H. Smith, J.J. Skehel, and M.J. Turner, pp. 101-114. Weinheim: Verlag Chemie.

(25) Rubin, D.H., and Fields, B.N. 1980. The molecular basis of reovirus virulence: The role of the M2 gene. J. Exp. Med. 152: 853-868.

(26) Stevens, J.G., and Cook, M.L. 1971. Latent herpes simplex virus in spinal ganglia of mice. Science 173: 843-845.

(27) Stevens, J.G.; Todaro, G.J.; and Fox, C.F. 1978. Persistent Viruses. New York: Academic Press.

(28) Tamm, I. 1975. Cell injury with viruses. Amer. J. Path. 81: 163-177.

(29) Trachsel, H.; Sonenberg, N.; Shatkin, A.J.; Rose, J.K.; Leong, K.; Bergman, J.E.; Gordon, J.; and Baltimore, D. 1980. Purification of a factor that restores translation of vesicular stomatitis virus mRNA in extracts from poliovirus-infected Hela cells. Proc. Natl. Acad. Sci. USA 77: 770-774.

(30) Wechsler, S.L.; Weiner, H.L.; and Fields, B.N. 1979. Immune response in subacute sclerosing panencephalitis: reduced antibody response to the matrix protein of measles virus. J. Immunol. 123: 884-889.

(31) Weiner, H.L.; Ault, K.A.; and Fields, B.N. 1980. Interaction of reovirus with cell surface receptors. I. Murine and human lymphocytes have a receptor for the hemagglutinin of reovirus type 3. J. Immunol. 124: 2143-2148.

(32) Weiner, H.L.; Drayna, D.; Averill, D.R.; and Fields, B.N. 1977. Molecular basis of reovirus virulence: Roles of the S1 gene. Proc. Natl. Acad. Sci. USA 74: 5744-5748.

(33) Weiner, H.L.; Powers, M.L.; and Fields, B.N. 1980. Reovirus virulence and central nervous system cell tropism: Absolute linkage to the viral hemagglutinin. J. Inf. Dis. 141: 609-616.

Neuronal-glial Cell Interrelationships, ed. T.A. Sears, pp. 229-250.
Dahlem Konferenzen 1982. Berlin, Heidelberg, New York: Springer-Verlag.

Mechanisms of Cell Injury by Certain Protein Toxins

A. M. Pappenheimer, Jr.
Biological Laboratories, Harvard University
Cambridge, MA 02138, USA

CONTENTS

INTRODUCTION

Polypeptide neurotoxins may be grouped, arbitrarily, into at
least four classes on the basis of their mode of action:

Class I: Toxins that effect Na^+ channels on excitable membranes
and which stimulate cAMP and/or cGMP formation. As examples,
we may cite scorpion toxins, toxins produced by certain sea
anemones, and other invertebrate venoms. Cholera toxin which
stimulates cAMP formation might also be included in this group
but will be considered under Class IV toxins.

Class II: Snake venoms that act postsynaptically by binding
with high affinity to ACh receptors to prevent them from interacting with ACh, thereby causing neuromuscular blockade and

paralysis. For many of these so-called "cardiotoxins" or "long chain venoms," such as α-bungarotoxin, the complete amino acid sequence of 71-74 residues with 5 disulfide bridges is known. Structurally related short chain neurotoxins from various sea snakes and other species with only 57-66 residues cause similar effects but have lower affinity for ACh receptors. Snake venoms also contain phospholipases A and C which cause release of lyso- lecithin which may act as a detergent to disrupt membrane struc- tures.

Class III: Toxins that selectively attack neuronal constituents and act presynaptically to prevent chemical transmitter release. Typical examples are tetanus toxin and the various botulinal toxin types A-F.

Class IV: Toxins which do not act selectively on nervous tissue, but may damage or kill nerve cells in culture or in animals if introduced directly into the PNS or CNS. If introduced into ani- mals by other routes in low dosage (as in diphtheritic neuropathy), these toxins may cause lesions in the PNS with accompanying de- layed reversible paralysis in survivors.

This review will be restricted to consideration of certain plant and microbial toxins that fall under Classes III and IV. These toxins, some properties of which are summarized in Tables 1 and 2, have been selected because they have been most carefully studied at both pharmacological and molecular levels.

STRUCTURE OF CLASS III AND IV TOXINS (7,9,26)
All of the protein toxins listed in Table 1 have been isolated in a purified homogeneous state and all but Pseudomonas exotoxin A and E. coli LT enterotoxin have been crystallized. The complete amino acid is known for ricin and for cholera toxin and has almost been completed in the case of diphtheria toxin.

All of these toxins have been shown to consist of two covalently linked, though functionally distinct, domains which, following the nomenclature first proposed by Olsnes and Pihl (25) for the

TABLE 1

PHYSICAL PROPERTIES OF SOME PROTEIN TOXINS

| TOXIN | CLASS | MOLECULAR WEIGHT | | | CHO |
		Toxin	Effectomer	Haptomer	
Tetanus	III	150,000	55,000	95,000	O
Botulinum (A-E)	III	160,000	55,000	105,000	O
Diphtheria	IV	61,000	21,150	ca40,000	O
Ps. exotoxin A	IV	66,000	ca26,000	ca40,000	O
Cholera toxin &	IV	84,000	26,500	57,500	O
E. coli LT toxin				(=5x11,500)	
Abrin	IV	65,000	30,000	35,000	+
Ricin	IV	66,000	32,000	34,000	+
Modeccin	IV	66,000	28,000	38,000	+

toxic plant lectins abrin and ricin, we may term "effectomer"
and "haptomer." Neither domain has significant toxicity if isolated free from the other. The effectomer (or A chain of the
toxic lectins, fragment A of the microbial toxins, L or light
chain in the case of tetanus and botulinum toxins) must reach
the cell cytoplasm in order to exert the lethal or other characteristic biological effect of the toxin. The haptomers (B chain
of the toxic lectins, fragment B of microbial toxins, or heavy
chain (H) in the case of tetanus toxin) contain determinants that
recognize and bind to specific receptors on the surface of target cells and in addition facilitate translocation of the effectomer across the plasma membrane to reach the cytoplasm. The
mechanism(s) by which this interesting and specific translocation
is accomplished is currently being investigated in many laboratories and will be discussed in some detail in a later section.
It is the haptomer that determines host specificity and distribution of lesions within the intoxicated host. Thus when toxins
with similar substrates at the molecular level, as is the case
with diphtheria toxin and Pseudomonas toxin A on the one hand
and with abrin, ricin, and modeccin on the other, are introduced
into a given animal species, the particular organs most affected
and the severity of the injury produced may be quite different
for each toxin.

MICROBIAL TOXIN PRODUCTION AND ITS GENETICS (23,33)

With the exception of E. coli enterotoxin, all of the bacterial toxins listed in Table 1 are secreted into the extracellular culture medium; in most cases during the terminal stages of bacterial growth. Diphtheria and cholera toxins are synthesized as "pretoxins" with a short hydrophobic N-terminal amino acid sequence that is split off during secretion and which appears to be analogous to the signal or leader peptides of eukaryotic secreted proteins. The synthesis of diphtheria toxin takes place on the cell membrane. Although direct evidence is lacking, it

Table 2 - Biological Properties of Some Protein Toxins.

TOXIN	ENZYMIC ACTIVITY OF		EFFECTOMER	RECEPTOR	BIOCHEMICAL and/or
	Target substrate	Cofactor	Product		PHARMACOLOGICAL EFFECTS
Botulinum (A-E)	?	?	?	?	Preferentially attacks PNS cholinergic synapses. Acts presynaptically to block ACh release.
Tetanus	?	?	?	ganglio-sides GD_{1b} & GT_1	Acts on both PNS & CNS presynaptically to block ACh release (PNS); glycine and GABA release (CNS).
Diphtheria toxin & Ps. exotoxin A	eEF2	NAD	ADPR-eEF2	?	Blocks protein synthesis in target cells. Necrosis and hemorrhage in affected tissues.
Cholera & E.coli LT enterotoxins	GTP binding subunit of AD-cyclase	NAD	ADP ribosylated subunit	GM_1	Activation of AD-cyclase causing elevated intracellular cAMP levels.
Abrin, ricin & modeccin	eukaryotic ribosomes	none	inactivated 60S ribosome	glycoprotein with terminal galactose	Inhibition of protein synthesis. Necrosis and hemorrhage in affected tissue.

seems likely that the other toxins secreted by the Gram$^+$ bacteria are also synthesized on membrane-bound polysomes.

The structural gene for diphtheria toxin (*tox*) is carried by a
temperate bacteriophage. Although only lysogenic bacteria have
been shown to produce toxin, the *tox* gene is not essential for
phage replication and is located on an operator-promoter region
of its own that apparently contains no other phage genes. The
location of the *tox* gene on the genetic map of toxigenic coryne-
phage β has been determined. Recently, a physical map of coryne-
phage β^{tox+} has been constructed and reported by two laboratories
working independently (Costa et al. and Buck & Groman in preparation). Both groups have isolated the *tox* gene operon on a small
ca 3000 base pair dsDNA restriction enzyme fragment.

Although the diphtheria *tox* structural gene is carried by a
phage genome, its expression is controlled by an iron-containing
bacterial host repressor protein that binds to the *tox* promoter
or operator but dissociates when iron becomes growth-limiting.
The normal function of this bacterial repressor protein, that is
present even in the sensitive uninfected host bacterium, is not
known.

The fact that *tox* is non-essential for phage replication has
greatly facilitated the isolation of mutant *tox* gene products
(CRM's) that are serologically related to toxin, but which are
either nontoxic or of greatly reduced toxicity. Thus when
β-phage are treated with a mutagenic agent at a concentration
sufficient to kill about 99.5% of the phage, a high percentage
of the survivors will carry mutations in the *tox* operon. Some
properties of a few such mutants which have proved most useful
in studies on the biochemistry of the toxin molecule and in
the mechanism by which its A fragment crosses the plasma membrane of susceptible mammalian cells are shown in Table 3.

The structural *tox* gene for E. coli LT toxin is present on a
large transmissible plasmid. The LT-DNA region of this plasmid

234

Table 3

Some properties of CRMs related to diphtheria toxin

Protein	Approximate mol wt	Toxicity[a] (MLD/μg)	Enzyme Activity[b] (%)	Binding[c]
Toxin	60,000	25-30	100	(+)
Toxoid	60,000	0	0	-
Fragment A	21,150	0	100	-
CRM45	45,000	0	100	-
CRM197	60,000	0	0	+
$A_{45}B_{197}$	60,000	25-30	100	(+)

[a]One MLD is the amount of toxin that will kill a 250g guinea pig in 4-5 days.
[b]ADP-ribosylating activity after "nicking" and reduction relative to fragment A as 100%.
[c]Ability to compete with toxin for cell membrane receptors.

has now been isolated as a fragment (≤ 1.2 x 10^6 daltons) which has now been joined to a cloning vector and cloned (8). The fragment has been shown to encode two proteins of 11,500 and 25,000 daltons respectively. The larger of the two proteins has been shown to have NAD-mediated AD-cyclase stimulating activity. The two cistrons appear to share a common promoter.

The *tox* gene for cholera toxin appears to be encoded by the V. cholera genome which has made isolation of CRM's difficult. Bacterial (*htx*) regulatory mutants have been isolated which produce high yields of toxin and the *htx* gene has been mapped on the V. cholera chromosome (18). Mutant strains that appear to produce only choleragenoid, i.e., the toxin B subunit, have been isolated. These strains fail to produce any material that cross-reacts with antibody against the A subunit although culture filtrates react strongly with anti-cholera B subunit serum (13). It has been shown that the cholera toxin A subunit is synthesized by free polysomes as a 45,000 dalton precursor (24).

Very little is known about the genetics of production of the
other toxins listed in Table 1. There is evidence that certain
types of botulinum toxins are only produced by lysogenic strains
of Clostridium botulinum but no CRM's have been reported. Very
recently, the *tox* gene for tetanus toxin has been shown to be
on a large plasmid.

ENZYMIC ACTIVITY OF CLASS IV EFFECTOMERS
Diphtheria toxin: Intact diphtheria toxin is enzymically inactive. Activation requires reduction of a cystine disulfide
at position 186 and cleavage of a peptide bond in the exposed
loop of 14 amino acids it subtends. The molecule is thereby
split into two fragments, A and B, which nevertheless remain
tightly held together by weak interactions despite a confor-
mational change. The split molecule retains its toxicity even
after carboxymethylation of the sulfhydryls, although neither
fragment is toxic when separated from the other. The effectomer, fragment A, catalyzes the NAD-mediated ADP-ribosylation
and inactivation of eEF2 in cell extracts from all eukaryotic
species thus far examined according to the following reaction:

$$NAD + eEF2 \rightleftharpoons ADP\text{-}eEF2 + nicotinamide + H^{+} \tag{1}$$

Some years ago Robinson et al. (29) sequenced a 15 amino acid
labeled polypeptide isolated from a tryptic digest of C^{14}-ADP
ribosylated eEF2 from rat liver. The labeled ADP ribosyl
group was linked to a basic amino acid which they were unable
to identify. Almost identical tryptic peptides have now been
isolated from P^{32}-ADP ribosylated eEF2's from yeast, wheat germ,
and bovine liver. The unusual amino acid has been identified
as a histidine residue containing a unique modification on car-
bon # 2 of the iminazole ring. The ADP ribose group attaches
to one of the two ring nitrogen atoms. The new amino acid, the
structure of which is given below, has been named "diphthamide"
(36).

$$
\begin{array}{ccc}
HC = C - CH_2CHCOOH \\
| \quad | \qquad\qquad | \\
HN \diagdown \quad N \qquad NH_2 \\
C \\
| \\
CH_2CH_2CHCONH_2 \\
| \\
N(CH_3)_3{}^+
\end{array}
$$

Diphthamide has not been found in prokaryotes nor in any eukaryotic protein other than eEF2. It is no wonder that reaction 1
is highly specific. Because of this extreme specificity, it has
been possible to develop rapid methods for assay of eEF2 in crude
tissue or cell extracts. Provided that a few simple precautions
are taken to avoid side reactions, the only TCA-precipitable labeled protein in such extracts, after addition of P^{32}-NAD and
fragment A, will be P^{32}-ADP ribosyl-EF2. Thus one can readily
follow the kinetics of entry of fragment A into intoxicated cells.

Despite the fact that diphthamide appears to be universally present in EF2 of all eukaryotic cells, mutants lacking the modification can be isolated from mammalian cell cultures with relative ease because they are completely resistant to diphtheria
toxin even in high concentration (22). It is of interest that
MOD^- mutants from cultured cells grow and synthesize protein at
the same rate as the wild type from which they were derived.
At present we have no clue as to the function of this ubiquitous
modified amino acid.

Pseudomonas exotoxin A. Although pseudomonas toxin is synthesized by an unrelated bacterial species and is unrelated serologically to diphtheria toxin, its mode of action is similar.
Like diphtheria toxin, its effectomer catalyzes the NAD mediated ADP-ribosylation of eukaryotic EF2 and MOD^- mammalian cell
lines are completely resistant to both toxins. Yet there is
no demonstrable immunological cross-reactivity between the two
toxins, either by neutralization of toxicity or of enzymic
activity. Activation of pseudomonas toxin requires a confor-
mational change which may result either from cleavage of a

peptide bond so as to release a 26,000 dalton polypeptide containing the toxin's N-terminus, or following reduction of the toxin with thiol in the presence of a denaturing agent such as urea (7).

TRANSLOCATION ACROSS THE CELL MEMBRANE

For all of the Class IV toxins listed in Table 1, the A chain or effectomer must traverse the plasma membrane and reach the cytoplasm in order to kill or injure its target cell. The mechanism by which large, charged polypeptide molecules are translocated across a lipid bilayer is a problem of considerable current interest. At least in the case of toxins, the translocation need not be a particularly efficient process, since only a few molecules need arrive at the cytoplasm to cause death or injury to the cell. Thus, at concentrations of NAD and eEF2 approximating those of living mammalian cells, the turnover of diphtherial fragment A is about 2000 molecules per minute at 37C. A simple calculation will show that at a steady-state level of 5 or 6 molecules of A per cell, 50% of its EF2 content will become ADP ribosylated within 2-3 hours. In fact Yamaizumi et al.(40) have devised a method for introducing small numbers of macromolecules into mouse L cells and then using a cell sorter to isolate those with a single molecule per cell. Cells into which one molecule of fragment A had been inserted were incapable of giving rise to colonies. It was also shown (Uchida, personal communication) that diphtherial fragment A is stable for at least 18 hours within the cytoplasm, whereas fragment B and various other proteins are rapidly degraded. There is evidence that NAD protects fragment A from hydrolysis by serine proteases. Finally, it was shown that the A fragment of nontoxic CRM197, a protein that does not interact with NAD because of a missense mutation which has rendered it enzymically inactive, is rapidly degraded when introduced into cells. Thus only the effectomer need traverse the membrane; if the whole toxin molecule reaches the cytoplasm its haptomer will be rapidly degraded.

Receptors: It is generally agreed that the first step in the
entry process is a reversible interaction between specific surface receptors on the target cell membrane and determinants located within the toxin's haptomer domain. For each of the three
toxic lectins, abrin, ricin, and modeccin, this interaction is
between the respective haptomers and membrane glycoproteins
with terminal nonreducing galactose residues. Treatment of cells
with neuraminidase to expose additional galactose residues increases sensitivity to these toxins; mutant cell lines with a
reduced number of exposed terminal galactose residues are relatively resistant. Galactose, lactose, and various glycoproteins
inhibit binding of the haptomers and reduce toxicity for cultured cells. However, there are normally a great many different membrane proteins with terminal galactose residues and
binding studies have shown that HeLa cells, for example, may
bind 10^7 or more ricin molecules per cell. It seems probable
that only a very small fraction of these binding sites can
actually function in the entry process. It has been shown that
abrin and ricin A-chains inactivate buffered suspensions of salt-
washed ribosomes at the rate of 1500 ribosomes per minute per A-
chain molecule. Like diphtherial fragment A, therefore, only a
very few effectomer molecules need reach the cytoplasm to kill
a cell (25).

The receptor for cholera and E. coli enterotoxins has been identified as the ganglioside GM-1 (9). Since this ganglioside is
present on the surface of all or almost all animal cells, these
enterotoxins, like the toxic lectins are relatively nonspecific
in host range and large numbers of toxin molecules can be bound
per cell. Free GM-1 inhibits cytotoxicity, whereas an increase
in cell surface GM-1 renders a cell more sensitive. Each of
the five subunits of haptomer will bind one molecule of GM-1.

Little is known of the nature of the specific receptors on membranes of mammalian cells that are sensitive to diphtheria toxin, although Chin and Simon (5) have isolated a protein fraction
from rabbit liver which, at a concentration of only a few ngs/ml,
will protect sensitive cells from intoxication. Whatever the

nature of this receptor may be, it is not ubiquitous in distribution as seems to be the case with receptors for the toxic
lectins and for cholera and tetanus toxins. Receptors for
diphtheria toxin have only been demonstrated on the surface of
cells from sensitive mammalian species. Most tissues and cell
lines from resistant species (e.g., mice and rats) do not possess exposed receptors, and there is no evidence for specific
receptors on cells of cold-blooded animals or invertebrates
even though they may be sensitive to high toxin concentrations.
Most sensitive cell lines bind less than 10-20,000 toxin molecules per cell, although African green monkey kidney cells
(CV, Vero), which are exceptionally sensitive, have about 10
times as many receptors per cell (20). From competition studies
using CRM197 and I^{125}-labeled toxin it has been shown that
binding constants for both HeLa and CV-1 cells are roughly the
same (K_A = ca 10^8L/mole), despite the fact that the former cell
line is about 100 times more resistant than the latter.

Although we know very little about the chemical nature of the
specific membrane receptors for diphtheria toxin, we do know
the approximate location of the receptor-recognition site on
the toxin molecule. CRM's 45 and 47, which lack the positively
charged 12-15,000 dalton C-terminal amino acid sequence of
intact toxin including the remaining cystine residue, do not
interact with membrane receptors and therefore cannot protect
sensitive cells from intoxication. It has been shown that
certain negatively charged small molecules such as nucleotide,
tri- and tetra-phosphates, inositiol hexaphosphate and various
inorganic polyphosphates can protect, competitively, sensitive
cells from wild type toxin (19). Thus ATP binds tightly to
what has been called the "P-site" which is located near the C-
terminus of the toxin haptomer; a site that is missing from
CRM45 (16). Recent evidence suggests that, spatially, the P-
site is located very close to the active NAD-binding site on
the effectomer. These studies have raised the possibility
that the initial reversible step in intoxication may involve
interaction between the diphtherial P-site and the cell surface receptor.

The entry process - translocation of effectomer: The second
step(s) in the translocation process involves irreversible
entry of all or part of the toxin molecule into the lipid bilayer of the plasma membrane. Any explanation of how this
translocation is effected must account for the fact that toxin
concentrations, so low that no more than 3 or 4 receptor sites/
cell will be occupied at any given time are still capable of
killing the cell. We shall confine our discussion of this
portion of the entry process mainly to a consideration of
diphtheria toxin.

Some years ago, Clarke (6) showed that almost all membrane
proteins bind considerable amounts of the H^3-labeled nonionic
detergent, Triton X-100, whereas cytoplasmic and extracellular
proteins bind no detergent at all, with the exception, of course,
of those involved in lipid transport. Native diphtheria toxin,
like other typical soluble proteins, binds no Triton. Denatured toxin, on the other hand, binds about 45 moles Triton
X-100 per mole, all of which is bound to that portion of the
haptomeric B fragment that is present on CRM45 (4). In fact,
B45 binds 1.3 times its weight of this detergent.

Recent studies on the amino acid sequence and circular di-
chroism of CNBr cleavage products isolated from diphtherial
fragment B have confirmed and extended these observations on
Triton X-100 binding. Lambotte et al. (15) found that most of the
α-helical structure of the toxin molecule is located within
that part of fragment B that is present in CRM45. There is an
amphipathic stretch of 17 amino acids near the N-terminus of
fragment B which is part of a sequence of some 60 amino acids
showing striking homology with human apolipoprotein Al. There
is also a hydrophobic α-helical stretch of 27 amino acids
which closely resembles the lipid-associating domains of intrinsic membrane proteins that are responsible for anchoring
them in the bilayer.

It is not surprising, therefore, that CRM45 and its B45 fragment readily insert into liposomes and membrane vesicles. It
has recently been shown that picomolar concentrations of CRM45,
B45, and _denatured_ toxin will, like colicin K (31), form channels at pH's below 6 in artificial lipid bilayers and unilam-
melar liposomes through which ions and simple sugars are freely
diffusible (B. Kagan and A. Finkelstein, personal communication).
The channels are about 18 Å in diameter and are certainly wide
enough to accommodate fragment A in its extended denatured
form. In the case of living cells, it is possible that channel
formation across the plasma membrane must be preceded by internalization of individual toxin molecules into small vesicles.
If the pH inside such vesicles falls below 5-6, i.e., low enough
to denature the toxin, channels will presumably be formed. It
has been well established (14) that NH_4CL and chloroquine, reagents that are known to raise th pH within endocytotic vesicles,
will protect cells from diphtherial intoxication. In contrast,
Sandvig and Olsnes (30) have recently obtained evidence that
brief exposure of HeLa cells to toxin at low pH increases their
sensitivity and counteracts the protective effect of NH_4CL.
If reducing conditions exist within such vesicles, and if nicking
of the exposed cys186 loop occurs, occasional molecules of fragment A may reach the cytosol by diffusion. It has been suggested
that the receptor itself may be the specific protease responsible
for nicking and possibly for cleavage of an additional peptide
bond near the second cystine so as to release the hydrophilic
positively charged C-terminal region of the toxin (27).

We may now summarize the proposed entry mechanism for diphtheria
toxin as follows: a) Reversible interaction with specific surface receptors on target cell. b) Internalization of individual
toxin molecules within small vesicles. c) Denaturation when pH
falls within the vesicle. d) Channel formation with concomitant
reduction of cys186 disulfide and activation of membrane protease
(receptor?) to cleave peptide bonds in exposed loop and to release positively charged C-terminal "tail." e) Diffusion of extended form of A fragment through channel and renaturation to
active effectomer in cytosol.

Other toxins: It is suggested that the effectomers of other
toxic proteins and perhaps of protein hormones and growth factors may reach their cytoplasmic target by mechanisms analogous
to that proposed for diphtheria toxin. In the case of cholera
toxin, Gill had already proposed (9) that the pentomeric B
subunits are arranged in a plane, as in a doughnut, with the
A subunit resting on top in the center. When the subunits
interact with GM-1 ganglioside on the cell surface, the "doughnut" may sink into the membrane and after reduction of the A
subunit disulfide bond, the A_1 effectomer is released to slip
through the channel opened up by the pentomeric structure.
Evidence that cholera toxin forms channels in artificial bilayers
containing GM-1 has been reported (35). In the case of ricin,
it has been shown that there is a major stretch of hydrophobic
amino acids near the C-terminus of the A chain, rather than
near the N-terminal of the haptomer as is the case with diphtheria toxin.

MODE OF ACTION OF BOTULINUM AND TETANUS TOXINS (2,10,32)
Virtually nothing is known at the molecular level about the
mode of action of Class III toxins. Both botulinum and tetanus toxins act presynaptically to block ACh release in the PNS.
However, in experiments with a mouse phrenic nerve-hemidiaphragm
preparation, botulinum neurotoxin was about 2000 times more po-
tent than tetanus toxin. On the other hand, only tetanus toxin
acts on the CNS where it has been shown to act presynaptically
at terminal synapses to block release of the transmitters,
glycine and GABA.

Since the early studies of Meyer and Ransome in 1903, a great
deal of interest has centered on how tetanus toxin is transported from the peripheral nerve endings to reach the CNS.
Recent experiments with I^{125}-labeled tetanus toxin have confirmed the early postulate that toxin is first bound to terminal neuromuscular junctions and is then carried to the CNS
by retrograde intra-axonal transport. In rodents, the speed
of transport has been calculated to be about 0.5-1 cm/hr.

The receptors for tetanus toxin appear to be the gangliosides GD_{1b} and GT_1. These bind tightly to sites located on the so-called fragment C, a protease cleavage product (Mr 47,000) of the toxin heavy chain which contains its C-terminus. Primary cultures of neuronal cells with exposed GD_{1b} and/or GT_1 on their surface will bind the toxin; cells lacking these ganglio-sides will not.

Nothing is known about the mechanism of retrograde intraaxonal transport of tetanus toxin. The studies with ^{125}I-labeled tox-in do not tell us whether the whole toxin molecule enters neuronal cells and travels up the axon to terminal synapses in the ventral horn of the CNS, or whether it is only the light or A chain, which in anology with other toxins has been assumed to carry the effectomer activity.

Recently, the action of tetanus and botulinum toxins has been studied in primary nerve cell cultures. Both toxins were shown to suppress ^{3}H-ACh release upon K^+ depolarization (1). It is hoped that further studies with neuronal cultures may provide a clue as to the underlying molecular mechanisms.

DIPHTHERITIC NEUROPATHY (12,37,38)
A frequent late symptom seen in patients recovering from diph-theria is peripheral paralysis, especially of the soft palate. Meyer, nearly 100 years ago, observed degeneration of the my-elin sheath and demyelination of nerve fibers in human cases at autopsy. Diphtheritic neuropathy has proved somewhat diffi-cult to reproduce experimentally in guinea pigs and rabbits but may be accomplished by injection of slightly underneutral-ized horse antitoxin-toxin complexes. Guinea pigs injected into the thigh muscle with such complexes become paralyzed in the injected limb within a few days. In young chicks, direct injection of 10 pgs or less of free toxin (10^8 molecules) into a sciatic nerve will produce functional disability within 8-10 days and 100 pgs causes, within 5 days, severe paralysis with striking demyelination starting at the nodes of Ranvier. It

was noted early that the axons remained uneffected and it was
suggested that Schwann cells might be especially sensitive to
the toxin. Electron microscopic examination of the lesions
showed that this is indeed the case. The studies of Weller
and Mellick (39) demonstrated that intoxicated Schwann cells
showed greatly increased lysosomal activity as evidenced by a
sharp rise in acid phosphatase even before there was appre-
ciable myelin damage. It seems clear that the myelin degeneration with the accompanying paralysis is a delayed secondary
effect caused by lytic enzymes released from damaged Schwann
cells. An analogous situation appears to follow injection of
toxin directly into the CNS or into the brain where demyelination is presumably secondary to injured oligodendroglia.

Not long after the discovery of diphtheria toxin, Roux and
Borrel showed that small amounts of toxin injected into the
brains of rats caused paralysis and death, even though rats
were known to resist very much larger doses given intravenous-
ly. A few years ago, we observed that a few μgs of toxin were
lethal if injected intraperitoneally into mice bearing Ehrlich
ascites tumors. While there was some regression of the tumor,
the mice became paralyzed and died. No lesions were observed
in any of the tissues. To our surprise we found that CRM45
was equally toxic even though CRM45 has no discernable effect
if injected subcutaneously into guinea pigs or young chicks in
doses up to one mg/kg. Yet only 2-5 μgs/kg CRM45 is lethal
if injected intracerebrally into young chicks or mice. Even
by this route, the toxin effectomer, fragment A, is without
effect in doses up to 1 mg/kg. If injected directly into a
chick sciatic nerve, a few ngs CRM45 will cause demyelination
and paralysis within 6-7 days similar to that caused by toxin
itself (28).

CRM45 also has cytotoxicity for cultured cells. The rate of
protein synthesis in cultured eukaryotic cells, even cells
from "toxin-resistant" species, is inhibited by at least 50%
after 24-48 hrs exposure to 10^{-7}M either toxin or CRM45.

Fragment A at 10^{-6}M has no effect whatever, even on African
green monkey kidney cells that are sensitive to 10^{-13}M toxin
under similar conditions. Rat Schwann cells, however, are a
striking exception and are about 50 times more sensitive to
both CRM45 and toxin than other adult rat cells. In fact,
neuronal cells can be cultured in the presence of CRM45 or
toxin concentrations that are lethal for Schwann cells (J.B.
Brockes, personal communication).

Presumably because of its lipid-binding properties, CRM45 is
able to insert into plasma membranes and form a few channels,
provided it is present in high enough concentration. It is
known that the lipid and glycoprotein composition of the membranes of myelin-producing cells differs from that of other
cell membranes. Perhaps these differences in membrane composition will account for their greater sensitivity to CRM45 and
to toxin.

There appear to have been few attempts to investigate the effects of other Class IV toxins when injected into animals by
the intracerebral route. It has been reported that introduction of one µg cholera toxin into the CNS of rats causes increased locomotor activity lasting for as long as two weeks
(21). In a recent paper, Strocchi et al. (34) found that ricin
is highly toxic for rats when injected into the brain. Doses
of less than 0.1 µg were lethal within a week and the animals
showed severe neurological symptoms before death. As expected, protein synthesis in the brain was inhibited. However,
no mention was made of the histology of the lesions.

TOXIC INJURY AND MULTIPLE SCLEROSIS (11,17)
Multiple sclerosis (MS) is a disease characterized by regions
(plaques) of demyelination within the CNS in which axon continuity is preserved. There is a disappearance of myelin-producing oligodendroglial cells from these lesions. In many
respects the lesions resemble those produced following injection of small amounts of diphtheria toxin into the CNS of cats,

or those in the PNS that were discussed in the preceding section. MS is also characterized by spontaneous remissions during
which myelin-producing cells reappear and there is remyelination.

In the preceding section we stressed the fact that in diphtherial
neuropathy of the PNS, myelin degeneration is a secondary phenomenon which follows primary severe damage to Schwann cells. We
assume that the primary injury caused by diphtheria toxin to
myelin-producing cells is arrest of polypeptide chain elongation by ADP-ribosylation of EF2 as is the case with all sensitive cells. The dying cells release lysosomal enzymes that proceed to digest their own myelin. It follows that any agent
that would cause damage to myelin-producing cells so as to
release their lysosomal enzymes might be expected to cause lesions similar to those seen in MS or diphtheritic neuropathy.
For example, it has been shown that injection of lysolecithin
into the CNS of rats will produce local areas of demyelination
while leaving the axons intact (3). It might be of interest
to find out whether the α-toxin of Cl. perfringens, which is
a phospholipase C, would have a similar effect if injected into
the CNS.

It has often been suggested that MS may have a viral etiology.
A number of viruses are known which block protein synthesis in
their host cell. Thus polio virus appears to inhibit initiation of host protein synthesis. In conclusion, it seems to me
that any virus that can be shown to specifically damage and
kill myelin-producing cells in the CNS would be a likely etio-
logical agent for MS irrespective of the particular mechanism
by which it injures these cells.

*Abbreviations — ACh, acetylcholine · GABA, gamma-aminobutyric
acid · CNS, central nervous system · PNS, peripheral
nervous system · CRM, crossreacting material · eEF2, eu-
caryotic polypeptidyl-tRNA translocase · ADcyclase
adenyl cyclase · MS, multiple sclerosis.*

Acknowledgement. Supported in part by Grant PCM 79-04914 from
the National Science Foundation to Harvard University.

REFERENCES

(1) Bigalke, H.; Dimfel, W.; and Haberman, E. 1978. Suppression of ^{3}H-Ach release from primary nerve cell cultures by tetanus and botulinum-A toxin. Naunyn-Schmiedeberg Arch. Pharm. 303: 133-138.

(2) Bizzini, B. 1979. Tetanus toxin. Microbiol. Reviews 43: 224-240.

(3) Blakemore, W.F. 1976. Invasion of Schwann cells into the spinal cord of the rat following local injections of lysolecithin. Neuropath. & Applied Neurobiol. 2: 21-39.

(4) Boquet, P.; Silverman, M.S.; Pappenheimer, A.M., Jr.; and Vernon, W.B. 1976. Binding of Triton X100 to diphtheria toxin, CRM45 and their fragments. Proc. Nat. Acad. Sci. 73: 4449-4453.

(5) Chin, D., and Simon, M. 1981. Studies of the mechanisms of diphtheria toxin entry into cells in tissue culture. In Receptor Mediated Binding and Internalization of Toxins and Hormones, eds. J. Middlebrook and L. Kohn. New York: Academic Press, in press.

(6) Clarke, S. 1975. Size and detergent binding of membrane proteins. J. Biol. Chem. 250: 5459-5469.

(7) Collier, J.J.; Gilliland, D.G.; and Lory, S. 1979. Structure-activity relationships in diphtheria toxin and exotoxin A from Pseudomonas aeruginosa. In Transmembrane Signaling, pp. 751-759. New York: Alan R. Liss.

(8) Dallas, W.S.; Gill, D.M.; and Falkow, S. 1979. Cistrons encoding E. coli heat-labile toxin. J. Bact. 139: 850-858.

(9) Gill, D.M. 1977. Mechanism of action of cholera toxin. Adv. Cyclic Nucleic Acid Res. 8: 85-118.

(10) Haberman, E. 1978. Tetanus. In Handbook Clinical Neurology: Infections of the Nervous System, eds. P.J. Vinken and G.W. Bruyn, vol. 33. Amsterdam: North Holland Publishing Co.

(11) Hallpike, J.F. 1972. Enzyme and Protein Changes in Myelin Breakdown and Multiple Sclerosis. Stuttgart: G. Fischer Verlag.

(12) Harrison, B.H.; McDonald, W.I.; and Ochoa, J. 1972. Central demyelination produced by diphtheria toxin: An electron microscope study. J. Neurol. Sci. 17: 281-291.

(13) Honda, T., and Finkelstein, R.A. 1979. Selection and characteristics of a V. cholerae mutant lacking the A (ADPribosylating) portion of cholera toxin. Proc. Nat. Acad. Sci. 76: 2052-2056.

(14) Kim, K., and Groman, N.B. 1965. In vitro inhibition of
     diphtheria toxin by ammonium salts. J. Bact. 90: 1152-1562.

(15) Lambotte, P.; Falmagne, P.; Capiau, C.; Zanen, J.;
     Ruysschaert, J.M.; and Dirkx, J. 1980. Primary struc-
     ture of diphtheria toxin fragment B: structural simi-
     larities with lipid-binding domains. J. Cell Biol. 87:
     837-840.

(16) Lory, S., and Collier, R.J. 1980. Diphtheria toxin: Nu-
     cleotide binding and toxin heterogeneity. Proc. Nat. Acad.
     Sci. 77: 267-271.

(17) McDonald, W.I. 1974. Pathophysiology in multiple sclero-
     sis. Brain 97: 179-196.

(18) Mekalanas, J.J.; Seiblett, R.D.; and Romig, W.R. 1979.
     Genetic mapping of toxin regulatory mutations in V. cholerae.
     J. Bact. 139: 859-865.

(19) Middlebrook, J.L., and Dorland, R.B. 1979. Protection of
     mammalian cells from diphtheria toxin by exogenous nucleo-
     tides. Canad. J. Microbiol. 25: 285-290.

(20) Middlebrook, J.L.; Dorland, R.B.; and Leppla, S.H.
     1978. Association of diphtheria toxin with Vero cells -
     demonstration of a receptor. J. Biol. Chem. 253: 7325-7330.

(21) Miller, R.J., and Kelly, P.H. 1975. Dopamine-like effects
     of cholera toxin in the central nervous system. Nature
     255: 163-166.

(22) Moehring, J.M.; Moehring, T.J.; and Danley, D.E. 1980.
     Post-translational modification of EF2 in diphtheria toxin-
     resistant CHO cells. Proc. Nat. Acad. Sci. 77: 1010-1014.

(23) Murphy, J.E. 1976. Structure activity relationships of
     diphtheria toxin. In Mechanisms in Bacterial Toxinology,
     ed. A.W. Bernheimer. New York: John Wiley & Sons.

(24) Nichols, J.C.; Tai, P.-C.; and Murphy, J.R. 1980. Evi-
     dence that cholera toxin is synthesized in precursor form
     on free polysomes in V. cholerae. J. Bact. 144: 518-523.

(25) Olsnes, S., and Pihl, A. 1980. Toxic lectins and related
     proteins. In The Molecular Actions of Toxins and Viruses,
     eds. P. Cohen and S. van Heyningen. Amsterdam: Elsevier.

(26) Pappenheimer, A.M., Jr. 1977. Diphtheria toxin. Ann.
     Rev. Biochem. 46: 69-94.

(27) Pappenheimer, A.M., Jr. 1979. Transport of diphtheria
     toxin A fragment across the plasma membrane. In Trans-
     membrane Signaling, pp. 669-674. New York: Alan R. Liss.

(28) Pappenheimer, A.M., Jr.; Harper, A.A.; Moyniham, M.; and
     Brockes, J.P. 1981. Effect of route of injection on
     toxicity of diphtheria toxin and related proteins for cer-
     tain animal species: Cytotoxicity for cultured cells.
     (Submitted).

(29) Robison, E.Y.; Hendriksen, O.; and Maxwell, E.S. 1974.
     Elongation Factors: Amino acid sequence at site of ADP-
     ribosylation. J. Biol. Chem. 249: 5088-5093.

(30) Sandvig, K., and Olsnes, S. 1980. Diphtheria toxin entry
     into cells is facilitated by low pH. J. Cell Biol. 87:
     828-832.

(31) Schein, S.; Kagan, B.L.; and Finkelstein, A. 1978. Colicin
     K acts by forming voltage-dependent channels in phospho-
     lipid bilayer membranes. Nature 276: 159-163.

(32) Simpson, L.L. 1977. Presynaptic actions of botulinum
     toxin and α-bungarotoxin. In Specificity and Action of
     Animal, Bacterial and Plant Toxins, ed. P. Cuatrecasas.
     London: Chapman and Hall.

(33) Singer, R.A. 1976. Lysogeny and Toxigeny in Coryn. diph-
     theriae. In Mechanisms in Bacterial Toxinology, ed. A.W.
     Bernheimer, pp. 1-30. New York: John Wiley & Sons.

(34) Strocchi, P.; Novello, F.; Montanaro, N.; and Stirpe, F.
     1979. Effect of intraventricularly injected ricin on pro-
     tein synthesis in rat brain. Neurochem. Research 4: 259-267.

(35) Tosteson, M.T., and Tosteson, D.C. 1978. Bilayers con-
     taining gangliosides develop channels when exposed to chol-
     era toxin. Nature 275: 142-144.

(36) Van Ness, B.G.; Howard, J.B.; and Bodley, J.W. 1980. ADP-
     ribosylation of EF2 by diphtheria toxin: isolation and prop-
     erties of the novel Ribosyl amino acid and its hydrolysis
     products. J. Biol. Chem. 255: 10710-10716.

(37) Webster, H.deF.; Spiro, D.; Waksman, B.; and Adams, R.D.
     1961. Schwann cell changes in guinea pig sciatic nerves
     during experimental diphtheritic neuritis. J. Neuropath.
     exp. Neurol. 20: 5-34.

(38) Weller, R.O. 1965. Diphtheritic neuropathy in the chicken:
     An electron-microscope study. J. Path. Bact. 89: 591-601.

(39) Weller, R.O., and Mellick, R.S. 1966. Acid phosphatase
     and lysosomal activity in diphtheritic neuropathy and
     Wallerian degeneration. Brit. J. Exp. Path. 47: 425-434.

(40) Yamaizumi, M.; Mekada, E.; Uchida, T.; and Okada, Y. 1978.
     One molecule of diphtheria toxin A introduced into a cell
     can kill the cell. Cell 15: 245-250.

Neuronal-glial Cell Interrelationships, ed. T.A. Sears, pp. 251-269.
Dahlem Konferenzen 1982. Berlin, Heidelberg, New York: Springer-Verlag.

Distribution and Functional Significance of Sodium and Potassium Channels in Normal and Acutely Demyelinated Mammalian Myelinated Nerve

J. M. Ritchie
Dept. of Pharmacology, Yale School of Medicine
New Haven, CT 06510, USA

Abstract. The density of sodium channels in the axonal membrane of normal mammalian nodes of Ranvier is much higher than
in any other excitable tissue. In contrast, there are few if
any potassium channels. However, experiments in which the
myelin is acutely removed show that the internodal axonal membrane, which lacks sodium channels, is rich in potassium channels. This complementary distribution may, it is speculated,
be related to the fact that the functional specialization of
the mammalian fiber for rapid passage of impulses (high sodium channel density and low potassium channel density at the
node) may require some stabilizing influence in the paranodal
region to prevent re-entrant type phenomena (hence the internodal potassium channels).

INTRODUCTION

Conduction of the nervous impulse in myelinated nerve fibers,
as in excitable tissue generally, depends absolutely on the
presence in the axonal membrane of sodium channels through
which flow the sodium ions that carry the current generating
the action potential. The membrane of excitable tissue usually also contains potassium channels, but these are less essential for the conduction process proper in the sense that conduction of the impulse, although modified, is well maintained
when these potassium channels have been blocked by tetraethylammonium ions (see, for example, (35)). The present review

deals with three general questions. First, what is the normal
distribution of sodium and potassium channels in mammalian myelinated nerve?

Second, what are the differences between nodal and internodal
axonal membrane that are revealed by acute demyelination procedures? And, finally, what is the functional significance of
the differences between nodal and internodal membrane that are
indeed found, and what is their relevance to disease?

THE DISTRIBUTION OF SODIUM CHANNELS IN NORMAL NODES OF RANVIER OF MAMMALIAN MYELINATED FIBERS

Although the toxicity of tetrodotoxin, which is responsible for
Fugu fish (i.e., puffer fish) poisoning, and saxitoxin, which
is responsible for paralytic shellfish poisoning, have long
been known, their mechanisms of action did not become clear
until the mid 1960s when voltage-clamp experiments showed that
these toxins selectively abolish the sodium currents that are
essential for propagation of impulses in excitable membranes.
This observation soon led to the idea that the toxins physical-
ly block the sodium channels, and hence suggested a method
for counting the latter by measuring the amount of toxin taken
up by a nerve when conduction block had just occurred. In the
original, now classical, experiments of Moore, Narahashi, and
Shaw (20), a bioassay method was used to determine the amount
of toxin taken up by the tissue. Nowadays tritium-labelled
tetrodotoxin (12) or saxitoxin (29) are usually used. Binding
studies with such radioactively labelled toxin to a wide variety
of muscles and nerve show a consistent picture (Fig. 1). There
is a linear non-specific component of binding together with a
saturable, hyperbolic, component of binding that represents
binding of the toxin to some site in the sodium channel. The
concentration of toxin at which half the latter sites are occupied, the equilibrium dissociation constant K, gives an in-
verse measure of the affinity of the toxin for the channels.
The maximum value of the saturable component, M, gives a measure of the total number of channels in the tissue since it is

clear (13,26) that just one toxin molecule is required to block
a single channel. From the area of axonal membrane (determined

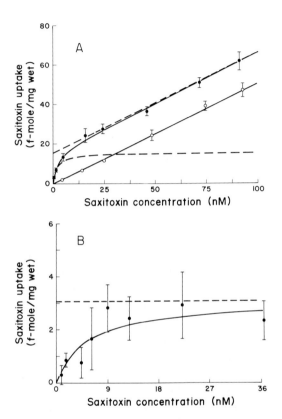

Fig. 1 - The uptake of labelled saxitoxin by the myelinated
fibers of the desheathed sciatic nerve. The upper panel shows
the total (●), linear (○), and saturable (interrupted line)
components of binding to intact rabbit sciatic nerve. The lower panel shows the saturable component of binding to the frog
sciatic nerve.

microscopically) per unit weight of a given tissue one can then
determine the channel density. These parameters are shown in
Table 1 for a variety of excitable tissues.

As can be seen in Table 1, for all tissues studied the equilibrium dissociation constant is small, being between 1 and 10 nM,
and, except for myelinated fibers to be discussed separately,
the channel density varies from 35 μm^{-2} for garfish olfactory
nerve to 550 μm^{-2} for the squid giant axon. These findings
thus confirm the original conclusion of Moore et al. (20) that
sodium channels are sparsely distributed over the membrane surface. For example, a channel density of 35 μm^{-2} corresponds
with one site for every 3,000,000 A^2 membrane. This means on
average that there is only one sodium channel for every 60,000
phospholipid molecules.

Plasticity
It should be noted that the channel density ($\underline{M}$) in any given
tissue is not an absolute, fixed, quantity, nor is the value
of $\underline{K}$. The channel density depends, for example, on the physiological state of the tissue. Thus, denervation of rat diaphragm muscle or of the extensor digitorum longis muscle leads
to a modest fall in the sodium channel density (2,10,27).
Furthermore, even under normal physiological conditions the
channel density may vary. For example, the channel density in
the sartorius muscle of a Southern variety of the frog Rana
pipiens is about twice that in the same muscle of a Northern
variety of the same species (27). Furthermore, variations can
occur between different muscles of the same animal. Thus in a
comparative study of the muscles of the hind limb of the rat,
Bay and Strichartz (2) found quite distinct differences in the
maximal saturable binding capacity of different types of muscle.
For example, in extensor digitorum longis (a fast twitch muscle)
the value of $\underline{M}$ was much higher than in soleus (a slow twitch
muscle); the value in the diaphragm was intermediate.

In addition to variations in $\underline{M}$, the value of $\underline{K}$ may also vary
depending on the prevailing conditions. Denervation produces
no significant change in $\underline{K}$ for muscle (2,10,27). However, the
value of $\underline{K}$ is sensitive to environmental factors such as temperature and ionic strength, which can alter $\underline{K}$ without affecting

$\underline{M}$ by modifying the effect of the fixed surface charges on the
nerve membrane. For example, Hille, Ritchie, and Strichartz
(14) showed in voltage-clamp experiments on frog myelinated
nerve that an increase in the concentration of calcium in the
bathing medium from 2 mM to 20 mM led to an 80% increase in the
value of $\underline{K}$ for the divalent saxitoxin and a 40% increase in the
value for the monovalent tetrodotoxin because it produces a
8 mV decrease in surface potential. It is not surprising, therefore, that the values of $\underline{K}$ reported in the literature differ
widely depending on whether it is measured in intact preparations
(where there is minimal disturbance of the phospholipid membrane

TABLE 1 - STX binding to excitable tissue

	Reference	$\underline{K}$ (n$\overline{M}$)	$\underline{M}$ (fmole.mg^{-1})	Channels per μm^2
Nerve				
Nonmyelinated				
Rabbit vagus	29	1.8	110	110
Lobster walking leg	29	8.5	94	90
Garfish olfactory	29	9.8	377	35
Squid (forbesi)	19	-	-	550
Squid (pealii)	36	3.8	-	170
Myelinated				
Frog sciatic	25	5.3	3.1	-
Rabbit sciatic	28	1.3	16	10,000
Rabbit optic*		3.7	15	-
Brain				
Rabbit*		6.0	136	-
Muscle				
Sartorius				
R. temporaria+	18	5.0	22	389
R. pipiens (S)	27	4.7	25	340
R. pipiens (N)	27	2.7	14	200
Diaphragm				
Rat	2	5.0	40	421
Rat	27	3.8	25	210
Rat denervated	27	4.1	16	-
Extensor Digitorum Longis				
Rat	2	4.3	54	535
Rat denervated	2	-	43	423
Soleus				
Rat	2	4.9	24	371

*Rang and Ritchie, unpublished data.   +Unlabelled TTX used.

structure), or in homogenized preparations (where vesicle formation and other changes in membrane conformation occur), or in
solubilized preparation (where the binding site may be partially
integrated into some detergent-produced micellar structure).

Optimum Density of Sodium Channels

On theoretical grounds Hodgkin (15) has calculated that there
is an optimum density of sodium channels at which conduction
velocity is maximal. This optimum occurs because increasing
the number of sodium channels not only increases the sodium conductance, thus tending to increase conduction velocity, but also increases that part of the membrane capacity associated with
the increased gating charge movement, which would tend to decrease conduction velocity. Calculation for the squid giant
axon (1,15) shows that conduction velocity should be maximal
when the density of sodium channels is 500-1,000 μm^{-2}, close
to that found experimentally in the squid in tetrodotoxin binding experiments (19).

In smaller diameter fibers, however, the density of channels is
found to fall well below this optimum: in the nonmyelinated
fibers of the rabbit vagus nerve where the average diameter is
0.7 μm, the channel density is 110 μm^{-2}; and in the garfish
olfactory nerve where the average fiber diameter is 0.24 μm,
the site density is only 35 μm^{-2}. There are two reasons why
the density of sites might have to be much less in small diameter
fibers. First, Hodgkin (15), has pointed out that spontaneous
initiation of impulses might be a frequent event if there were
500 channels per square micron in a 0.1 mm diameter axon with
a resistance of about 10^{10} ohm between inside and outside. A second, perhaps more cogent reason, concerns the expenditure of
energy required to keep pace with nervous activity. Because smaller
nerve fibers have a larger surface/volume ratio, their ionic
fluxes with each impulse would be larger (per unit weight of
tissue) if they had the same channel density as the squid. They
would thus have a correspondingly larger expenditure of energy
to maintain their resting condition and to restore the ionic

balance after each impulse. In fact, presumably as a result of
some process of adaptation involving in part a reduction in
channel density, both the ionic exchange associated with each
impulse and the resting and stimulated oxygen consumptions are
smaller in the garfish than in the vagus nerve, and the potassium loss in both is a good deal smaller than in the squid (24,30).

It should be noted that this argument for an optimum density for
maximum conduction velocity applies only to the case of non-
myelinated fibers where conduction is continuous. It would not apply to the case of myelinated fibers, where conduction is saltatory. Nor would the argument based on energy requirements apply
as forcefully (because only 1/1000 of the membrane is active).
And, indeed, as will be seen later, the channel density in myelinated fibers is a good deal higher in myelinated fibers than
in the squid giant axon.

Channel Density in Myelinated Fibers from Saxitoxin Binding
Experiments

Intact nerve. As Table 1 indicates, in most excitable tissue
the channel density is small - perhaps too small to be detected
by the electron microscope in freeze-fracture studies. However, the nodal membrane of the myelinated fiber seems to be
an exception in which the special electrophysiological characteristics (to give high conduction velocity) may require an unusual physical arrangement. For the ionic currents across nodal
membrane are much higher than in other excitable tissue: whereas
the peak sodium current density in the squid giant axon is usually about 1 mA.cm^{-2} and in frog skeletal muscle 2-4 mA.cm^{-2},
in the frog node of Ranvier it is nearly 2 orders of magnitude
greater being about 110 mA.cm^{-2} (21). This means that the sodium channel density at the frog node of Ranvier must also be
nearly 2 orders of magnitude greater than these other tissues.

Homogenized nerve. In intact nerve only the membrane at the
nodes is exposed; for the greater part of the axonal membrane
(perhaps about 1,000 times as much as at the node) is in the

internodal region where it is not normally exposed to the toxin
because it is protected by the myelin. This protection, however, can be removed by homogenizing the nerve so that all the
membrane, both nodal and internodal, is exposed. A surprising
finding is that where this is done, in spite of the fact that
the membrane exposed is increased by about 3 orders of magnitude,
the maximal binding capacity is the same as in the intact nerve
(26,28). This seems to mean that all the sodium channels are
located at the node of Ranvier with virtually no channels in the
internodal region. A similar biochemical inhomogeneity between
nodal and internodal membrane had already been noted as far as
staining with ferric iron and ferrocyanide is concerned (22,39).

Calculation of the Channel Density in Normal Mammalian and
Amphibian Nerves

As Figure 1 shows, the maximum saturable binding capacity for
saxitoxin in the myelinated fibers of the frog (3 f-mole.mg wet^{-1})
and the rabbit (17 f-mole.mg wet $^{-1}$) after correction for the
presence of non-myelinated fibers is about the same size as that
in muscle and a good deal smaller than that of nonmyelinated
nerves (see Table 1). Nevertheless, because of the restriction
of the channels to the node (revealed by the homogenization experiments), the actual channel density is extremely high. From
an electron microscopic study (40) of the spectrum of fiber
diameters in rabbit sciatic nerve, Ritchie and Rogart (28) calculated that there are about 800,000 μm^2 of nodal membrane per
mg wet nerve.

On the basis that all the sodium channels are located at the node
of Ranvier, the channel density at the rabbit node is therefore
about 10,000 μm^{-2} (25), i.e., 20-50 times higher than in muscle
or squid giant axon as indeed was predicted on the basis of the
electrophysiological experiments on the maximum sodium current
density. Furthermore, since the frog sciatic nerve binds only
about 1/5 as much saxitoxin as that of the rabbit, the channel
density at the frog node would be about 2,000 μm^{-2} (assuming
the spectrum of fiber diameters and other morphological characteristics of the frog sciatic are the same as in the rabbit).

This value agrees well with independent estimates of the channel density in frog nodes obtained from noise measurements and from gating current measurements (6,11,21), which give values of 1,300-5,000 μm^{-2}. Furthermore, it corresponds well with the number of 1200 μm^{-2} found in freeze-fracture studies of particles at the frog node (32).

Are There Sodium Channels in Schwann Cell Membrane?

One possible explanation for the calculated high channel density in the rabbit nerve might be that there are extranodal saxitoxin binding sites in the Schwann cell membrane, as has indeed been suggested for the squid giant axon (38). However, recent unpublished experiments (in collaboration with R. Pellegrino) seem to exclude this possibility. For within three days of cutting the cat sciatic nerve, the saxitoxin binding capacity distal to the cut begins to decrease progressively, falling to about 20% in seven days and to virtually zero by 24-49 days. Morphological evidence shows that from about 7 days onwards little of the axon remains, the space vacated having been filled by a vast proliferation of Schwann cells. Thus the disappearance of saxitoxin binding capacity in the face of the several-fold increase in both the number and area of Schwann cells suggests that in the normal nerve the bulk or all of the saxitoxin binding is to axonal membrane with little if any to Schwann cell membrane. A similar absence of saxitoxin binding sites had already been demonstrated for the glial cells of the optic nerve of the mud-puppy, Necturus Maculosa (37).

Electrophysiological Determination of Sodium Channel Density in Myelinated Fibers

The binding experiments seemingly predict that the sodium conductance of the frog node is 1/5 that of the rabbit. However, direct measurement in voltage-clamp records (9) shows that the conductances of rabbit and frog nodes are about the same. Furthermore, by measuring the gating currents at the same time as the sodium conductances, Chiu (6) showed that the number of sodium channels was also the same in both rabbit and frog node,

both being about 83,000 with a single channel conductance of
about 10pS. A nodal area of 66 μm^2 fiber (see below) would thus
give a channel density of about 1,300 μm^{-2}.

Is There Really a Discrepancy Between the Electrophysiological and Chemical Estimates of Sodium Channel Density?

At first sight there appears to be a large discrepancy between
the electrophysiological estimates of sodium channel density
and those determined by saxitoxin binding. However, it should
be emphasized that the electrophysiological calculations are
based on observations made on single nodes of Ranvier - usually
on the largest fibers of the nerve - whereas in the binding experiments the calculation applies to all fibers in a given nerve.
Furthermore, a fundamental assumption - that the spectrum of
fiber diameters and their distribution are similar in both nerves -
is untested. Finally, there is considerable uncertainty as to
the absolute value for the nodal area. Electrophysiologists
usually assume that a 15 μm diameter fiber has a nodal area of
66 μm^2 (34). However, the electrical capacity of the node,
which is about 2 pF in both kinds of fiber (6), would indicate
an effective nodal area of about 200 μm^2 if it were derived only
from a nodal membrane with a specific capacitance of 1 $\mu F.cm^{-2}$.
But direct electron microscopy gives the value of about 20
square microns for the rat node (3) and about 1/3 of that for
the frog node (31). No estimate of channel density at the node
can therefore be regarded at the moment as being completely reliable, so that the discrepancy is not yet firmly established.

THE INHOMOGENEOUS DISTRIBUTION OF SODIUM AND POTASSIUM CHANNELS ALONG THE NERVE MEMBRANE

The electrophysiological experiments agree with the saxitoxin
binding experiments in pointing to a very high density of sodium
channels in the nodal membrane. They (5,9) also confirm the
observation, first made on 1968 by Horackova, Nonner, and Stämpfli
(16), that the phase of late outward current carried by potassium
ions in frog and squid nerve is virtually absent in voltage-clamp
mammalian nodes of Ranvier (Fig. 2A). But certain pathophysiological findings suggest that mammalian nerve does not absolutely

lack voltage-dependent potassium channels. For example, large
outward potassium current, normally absent in rats, occur in
myelinated nerve fibers from alloxan diabetic rats (4). Furthermore, Sherratt, Bostock, and Sears (33) found that continuous
conduction can occur in demyelinated rat nerve fibers, and that
TEA and 4-aminopyridine, which are known to block potassium
channels, prolong the action potentials in this situation. Finally, Bostock and Smith (unpublished) have shown that within
2 days of exposure to lysolecithin some rat nodes exhibit a
phase of outward current that is sensitive to 4-aminopyridine.
These observations mean that either chronic disruption of the
myelin somehow, in time, induces formation of potassium channels
where they did not exist previously, or potassium channels are
normally present in parts of the axonal membrane hidden under-
neath the myelin.

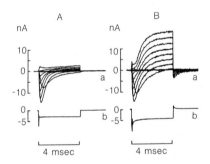

Fig. 2 - Ionic currents in rabbit node before (Aa) and after (Ba)
acute treatment to disrupt the myelin. Families of current in
both A and B were generated by the same series of depolarizations
from an initial holding value of -80 mV to various test potentials
in the range of -72.5 mV to +62.5 mV in 15 mV increments. The
current response associated with a hyper-polarization to -125 mV
is shown in Ab and Bb. The currents in A were measured in normal
Locke after 30 min of myelin-loosening treatment. During this
period, the outward current and the nodal capacity showed practically no change. Currents in B were measured 3 min after A
with the fiber in normal Locke when the late outward current (Ba)
and capacitative transient current (Bb) suddenly markedly increased.

Direct support for this latter possibility is provided by voltage-clamp experiments with rabbit node after acute treatment
designed to loosen the myelin from paranodal region (7,8). Such
treatment included application of various combinations of: lysolecithin (to dissolve the myelin); collagenase (to loosen the
connective tissue in the nodal region); high potassium Locke solution, hyper and hypo tonic solutions (to induce axonal volume
changes). At a critical stage during such treatment (Fig. 2Ba)
a large outward current suddenly appears which has all the characteristics of a potassium current: it is abolished by TEA and
caesium ions, has a reversal potential that becomes more positive when the external potassium concentration is increased,
and is kinetically similar to the known potassium current in
frog nodes. Furthermore, with its appearance comes the appearance of a slow transient capacitive component in the leakage
current (Fig. 2Bb) indicating up to a 60-fold increase in 'nodal' capacity. A plausible explanation is that this slow transient capacity current reflects charging of newly exposed axonal
membrane, probably in the paranodal region, which is uncovered
by the various acute demyelination treatments and that this exposed membrane contains potassium channels. This would account
for the occasional observation in newly dissected fibers of
large potassium currents even without treatment. If they are
deliberately stretched during dissection, a marked slow capacity
transient current is again consistently present.

At the time when the large outward current appears there is
never any substantial change in the size of the measured inward
sodium current, thus confirming the conclusion of Ritchie and
Rogart (28) from the saxitoxin binding experiments that there
are few if any sodium channels in the internodal region. Some
experiments were done to test directly the hypothesis that potassium but not sodium channels are normally present in the
internodal axonal membrane (Chiu and Ritchie, unpublished). A
stretch of internode, with no node, was mounted in the voltage-
clamp chamber. No ionic currents (other than leakage current),
of course, flowed (Fig. 3A) in response to a series of voltage
steps similar to those used to generate the currents in Fig. 2.

Nor were any currents present 90 min after exposure to lysolecithin (Fig. 3B). Suddenly, however, presumably as the myelin insulation was finally breached, a large outward current
appeared (Fig. 3C). No inward sodium current was observed.

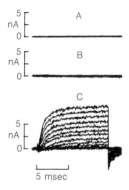

Fig. 3 - The effect of lysolecithin on the ionic currents
across the internodal membrane of a single rabbit sciatic
fiber. A small segment of internode (about 60 μm in length)
was voltage-clamped and a series of depolarizations, starting from a holding potential of -80 mV and ending at +70 mV
in 15 mV increments, was used to generate the currents. All
currents shown are corrected for a linear leakage. A shows
the control currents before lysolecithin treatment. B shows
the currents 90 minutes after an initial treatment of 3 minutes with lysolecithin (0.1%). C, taken 10 minutes after B,
shows the appearance of a large late current.

In summary, therefore, these experiments on acutely demyelinated
mammalian nerve demonstrate a complementary distribution of
sodium and potassium channels: the nodal membrane is rich in
sodium channels but poor in potassium channels, whereas the
internodal membrane is rich in potassium channels and had relatively few sodium channels.

DISTRIBUTION OF SODIUM AND POTASSIUM CHANNELS: THE NORMAL
FUNCTION AND RELATIONSHIP TO DISEASE
What is the function of the complementary distribution of so-
dium and potassium channels in mammalian myelinated nerve?

One reason for the absence of potassium channels from the node
may be lack of space. For the fast speed of conduction for
which myelinated fibers are specialized require a high nodal
sodium conductance, which entails a high density of sodium channels, and without potassium channels there would be more room
on the restricted nodal area for sodium channels. In addition,
there are two disadvantages to having potassium channels in the
node. First, during any prolonged increase in potassium conductance following an action potential (as in squid nerve), there
must be a temporary rise in the threshold current required to
excite the fiber again, and second, potassium current might lead
to an accumulation of potassium in the restricted nodal periaxonal space. For both reasons the ability of the node to generate an impulse soon after a preceding one would be compromised. Removal of potassium channels from the node would thus
increase the rate at which impulses would be passed along the
nerve. Clearly, there would be no need to remove potassium
channels from the internodal axonal membrane; and this may well
be why they persist there. However, it is interesting to speculate that they do have a positive role to play.

In normal mammalian myelinated nerve, delayed outward potassium current is absent (Fig. 2A) and so repolarization necessarily follows the inactivation of the sodium current. Yet
in spite of the absence of potassium currents the mammalian
action potential is still brief. In fact it is of somewhat
shorter duration than that in the frog under corresponding conditions of diameter, temperature, and potential (9), which means
that soon after the membrane potential has returned to its resting value the mammalian node is ready to respond to the next impulse. However, the danger is that any persisting depolarization in a patch of tissue near the node may re-excite the node,
causing repetitive firing after a single impulse analogous to
the re-entrant and ectopic phenomena well-known in cardiac muscle. The ring of juxtanodal axonal membrane in the paranodal
region might well be the locus of such a delayed, or slowly
developing, region of depolarization (because of the falling

off of sodium channel density and of the high series resistance of the external path between it and the node proper along
the periaxonal space), which in turn might re-excite the node
itself. On this basis the function of the potassium channels
in the paranodal region would be to dampen the paranodal response, and so prevent re-excitation. Indeed, preliminary computer simulation studies based on a high series resistance in
the region of the paranodal seal followed by a much lower series
resistance along the rest of the internode, and on a high axonal
membrane resistance, show that such re-excitation is feasible.
Too little, however, is known about the electrical parameters
of the internode to warrant further detailed speculation.

This stabilizing effect of potassium channels might also be
important in pathophysiological situations. Simply uncovering
potassium channels by lifting the myelin might be expected to
stabilize a mammalian nodal membrane. But pathological demyelination is clearly more complex, and in fact chronically demyelinated fibers (17) and genetically amyelinated fibers (23)
are hyper- rather than hypo-excitable, showing spontaneous
generation of impulses and repetitive firing. Such a phenomenon
may well be responsible for the paresthesias and the paroxysmal
symptoms of multiple sclerosis (23). One function of the potassium channels in the internodal region, particularly in the
paranodal region, therefore, might be to provide a chemical
voltage clamp to the paranodal area. The result would be to
offset the hyper-excitability of the damaged fiber by restrict-
ing the degree which a partially demyelinated patch of nerve
could generate a slowly developing depolarization, which might
subsequently reexcite other areas of the fiber.

CONCLUSION
In mammalian myelinated fibers there is a complementary distribution of sodium and potassium channels. The nodal membrane is
rich in sodium channels but contains relatively few potassium
channels. Acute demyelination experiments, however, show that,
in contrast, the internodal membrane does contain potassium
channels but relatively few, if any, sodium channels.

Acknowledgement. Supported in part by grant RG1162 from the National Multiple Sclerosis Society and by grants NS12327 and NS08304 from the USPHS.

REFERENCES

(1) Adrian, R.S. 1975. Conduction velocity and gating cur-
     rent in the squid giant axon. Proc. R. Soc. (London) B189:
     81-86.

(2) Bay, C.H., and Strichartz, G.R. 1980. Saxitoxin binding
     to sodium channels of rat skeletal muscle. J. Physiol.,
     in press.

(3) Berthold, C.H. 1978. Morphology of normal peripheral
     axons. In Physiology and Pathology of Axons, ed. S.G.
     Waxman, pp. 3-63. New York: Raven Press.

(4) Brismar, T. 1979. Potential clamp analysis on myelinated
     nerve fibres from alloxan diabetic rats. Acta physiol.
     scand. 105: 384-386.

(5) Brismar, T. 1980. Potential clamp analysis of membrane
     currents in rat myelinated nerve fibres. J. Physiol.
     298: 171-184.

(6) Chiu, S.Y. 1980. Asymmetry currents in the mammalian
     myelinated nerve. J. Physiol., in press.

(7) Chiu, S.Y., and Ritchie, J.M. 1980. Potassium channels
     in nodal and internodal axonal membrane of mammalian
     myelinated fibres. Nature 284: 170-171.

(8) Chiu, S.Y., and Ritchie, J.M. 1980. Potassium channels
     in the paranodal region of acutely demyelinated voltage
     clamped mammalian myelinated nerve. J. Physiol., in press.

(9) Chiu, S.Y.; Ritchie, J.M.; Rogart, R.B.; and Stagg, D.
     1979. A quantitative description of membrane currents
     in rabbit myelinated nerve. J. Physiol. 292: 149-166.

(10) Colquhoun, D; Rang, H.P.; and Ritchie, J.M. 1974. The
     binding of tetrodotoxin and α-bungarotoxin to normal and
     denervated mammalian muscle. J. Physiol. 240: 199-226.

(11) Conti, F.; Hille, B.; Neumcke, B.; Nonner, W.; and Stämpfli,
     R. 1976. Conductance of the sodium channel in myelinated
     nerve fibres with modified sodium inactivation. J. Physiol.
     262: 729-742.

(12) Hafemann, D.R. 1972. Binding of radioactive tetrodotoxin
     to nerve membrane preparations. Biochim. Biophys. Acta
     226: 548-556.

(13) Hille, B. 1968. Pharmacological modifications of the
      sodium channels of frog nerve. J. Gen. Physiol. 51: 199-219.

(14) Hille, B.; Ritchie, J.M.; and Strichartz, G. 1975. The
      effect of surface charge on the nerve membrane on the
      action of tetrodotoxin and saxitoxin in frog myelinated
      nerve. J. Physiol. 250: 34-45P.

(15) Hodgkin, A. 1975. The optimum density of sodium channel
      in an unmyelinated nerve. Phil. Trans. R. Soc. (London)B
      270: 297-300.

(16) Horackova, M.; Nonner, W.; and Stämpfli, R. 1968. Ac-
      tion potentials and voltage clamp currents of single rat
      Ranvier nodes. Proc. int. Union Physiol. Sci. 7: 198.

(17) Howe, J.F.; Calvin, W.H.; and Loeser, J.D. 1976. Im-
      pulses reflected from dorsal root ganglia and from focal
      nerve injuries. Brain Res. 116: 139-144.

(18) Jaimovich, E.; Venosa, R.A.; Shrager, P.; and Horowicz, P.
      1976. The density and distribution of tetrodotoxin re-
      ceptors in normal and detubulated frog sartorius muscle.
      J. Gen. Physiol. 67: 399-416.

(19) Levinson, S.R., and Meves, H. 1975. The binding of
      tritiated tetrodotoxin to squid giant axons. Phil.
      Trans. R. Soc. (London)B 270: 349-352.

(20) Moore, J.W.; Narahashi, T.; and Shaw, T.I. 1967. An
      upper limit to the number of sodium channels in nerve
      membrane? J. Physiol. 188: 99-105.

(21) Nonner, W.; Rojas, E.; and Stämpfli, R. 1975. Displace-
      ment currents in the node of Ranvier. Pfluegers Arch.
      354: 1-18.

(22) Quick, D.C., and Waxman, S.G. 1977. Specific staining
      of the axon membrane at nodes of Ranvier with ferric ion
      and ferrocyanide. J. Neurol. Sci. 31: 1-11.

(23) Rasminsky, M. 1978. Ectopic generation of impulses and
      cross-talk in spinal nerve roots of "dystrophic" mouse.
      Ann. Neurol. 3: 351-357.

(24) Ritchie, J.M. 1973. Energetic aspects of nerve conduc-
      tion: the relationships between heat production, electrical
      activity and metabolism. Prog. Biophys. Molec. Biol. 26:
      147-187.

(25) Ritchie, J.S. 1978. Sodium channel as a drug receptor.
      In Cell Membrane Receptors for Drugs and Hormones. A
      Multidisciplinary Approach, eds. R.W. Straub and L. Bolis,
      pp. 227-242. New York: Raven Press.

(26) Ritchie, J.M., and Rogart, R.B.   1977.   The binding of
     saxitoxin and tetrodotoxin to excitable tissue.   Rev. Physiol.
     Biochem. Pharmacol. 79: 1-50.

(27) Ritchie, J.M., and Rogart, R.B.   1977.   The binding of
     labelled saxitoxin to the sodium channels in normal and
     denervated mammalian muscle, and in amphibian muscle.   J.
     Physiol. 269: 341-354.

(28) Ritchie, J.M., and Rogart, R.B.   1977.   The density of so-
     dium channels in mammalian myelinated nerve fibers and the
     nature of the axonal membrane under the myelin sheath.   Proc.
     Nat. Acad. Sci. 74: 211-215.

(29) Ritchie, J.M.; Rogart, R.B.; and Strichartz, G.   1976.   A
     new method for labelling saxitoxin and its binding to non-
     myelinated fibres of the rabbit vagus, lobster walking
     leg and garfish olfactory nerve.   J. Physiol. 261: 477-494.

(30) Ritchie, J.M., and Straub, R.W.   1979.   Phosphate efflux
     and oxygen consumption in small non-myelinated nerve fibres
     at rest and during activity.   J. Physiol. 287: 315-327.

(31) Robertson, J.D.   1959.   Preliminary observations on the
     ultrastructure of nodes of Ranvier.   Zeitschrift fur Zell-
     forschung und Mikroskopische Anatomie 50: 553-560.

(32) Rosenbluth, J.   1976.   Intramembranous particle distribution
     at the node of Ranvier and adjacent axolemma in myelinated
     axons of the frog brain.   J. Neurocytol. 5: 731-745.

(33) Sherratt, R.M.; Bostock, H.; and Sears. T.A.   1980.   Effects
     of 4-aminopyridine on normal and demyelinated nerve fibres.
     Nature 283: 570-572.

(34) Stämpfli, R.   1954.   Saltatory conduction in nerve.   Physiol.
     Rev. 34: 101-112.

(35) Stämpfli, R., and Hille, B.   1976.   Electrophysiology of the
     peripheral myelinated nerve.   In Frog Neurobiology, eds.
     R. Llinas and W. Precht, pp. 1-32.   Berlin: Springer-Verlag.

(36) Strichartz, G.R.; Rogart, R.B.; and Ritchie, J.M.   1979.
     Binding of radioactively labelled saxitoxin to the squid
     giant axon.   J. Memb. Biol. 48: 357-364.

(37) Tang, C.M.; Strichartz, G.R.; and Orkand, R.K.   1980.
     Sodium channels in axons and glial cells of the optic nerve
     of Necturus Maculosa.   J. Gen. Physiol., in press.

(38) Villegas, J.; Sevcik, C.; Barnola, F.V.; and Villegas, R.
     1976.   Grayanotoxin, veratrine, and tetrodotoxin-sensitive
     sodium pathways in the Schwann cell membrane of squid nerve
     fibers.   J. Gen. Physiol. 67: 369-380.

(39) Waxman, S.G., and Quick, D.C. 1978. Intra-axonal ferric ion - ferrocyanide staining of nodes of Ranvier and initial segments in central myelinated fibers. Brain Res. 144: 1-10.

(40) Yates, A.J.; Bouchard, J.P.; and Wherrett, J.R. 1976. Relation of axon membrane to myelin membrane in sciatic nerve during development: comparison of morphological and chemical parameters. Brain Res. 104: 261-271.

Group on Injury

Standing, left to right: Klaus Felgenhauer, Cedric Mims,
Richard Johnson, Bernie Fields, Volker Ter Meulen, Barry Arnason,
Bertil Hille, Howard Weiner, Walter Gerhard.
Seated: Murdoch Ritchie, Byron Waksman, Tom Sears, Helmut Bauer,
Alwin Pappenheimer, Jr.

Neuronal-glial Cell Interrelationships, ed. T.A. Sears, pp. 271-286.
Dahlem Konferenzen 1982. Berlin, Heidelberg, New York: Springer-Verlag.

Injury
State of the Art Report

H. L. Weiner, Rapporteur
B. G. W. Arnason, H. J. Bauer, K. M. A. Felgenhauer,
B. N. Fields, W. U. Gerhard, B. Hille, R. T. Johnson,
C. Mims, J. M. Ritchie, A. M. Pappenheimer, Jr.,
T. A. Sears, V. Ter Meulen, B. H. Waksman

INTRODUCTION

One of the main themes to emerge from the group's discussion
of injury was that of specificity. The specificity of injury
to the nervous system in multiple sclerosis has not yet been
defined in terms of specific immune or viral-related damage
to either oligodendrocyte or myelin. Given a large body of
evidence suggesting that multiple sclerosis is an autoimmune
disease, the antigen against which the autoimmune response is
directed is yet to be defined. Similarly, viral induced injury
must also be understood in terms of mechanisms by which specificity of injury occurs. In some toxin models, excellent examples
of specificity of injury exist. (A. Pappenheimer, this volume).

The four areas of the group discussion were: a) physiology of
demyelinated nerve, b) mechanisms of immune injury, c) mechanisms of viral injury, and d) consequences of injury in relation to multiple sclerosis. Questions in these areas that were
raised during the course of the group discussion and were felt
by the group to be either major unanswered questions or questions that proposed new or novel ways of thinking about an

old problem have been integrated into the text. (The majority
of literature references are contained in the background papers
in this volume).

PHYSIOLOGY OF DEMYELINATED NERVE   (see Ritchie and Bostock and
McDonald, this volume).
A basic understanding of the physiology of conduction in myelinated nerve fibers and of the changes that occur following demyelination is necessary to understand how the injury of demyelination in multiple sclerosis ultimately affects neurologic function. This is important because the assessment of
the course of this disease is based mainly on the clinical assessment of function.

The myelin sheath surrounding an axon creates a low electrical capacitance and high resistance in the internode. Capacitance determines the number of ions needed to change the membrane potential,
resistance determines how rapidly charges "leak out" from the
nerve fiber. This optimizes the internode for the passive
spread of depolarization from an excited node of Ranvier to
bring the next node to its firing threshold and accounts for
the high speed of propagation of impulses in myelinated nerve.
In terms of interruption of function, anesthetizing a single
node of Ranvier does not block conduction of an impulse; however,
removal of myelin from an entire internode will stop conduction.

Interest in demyelinating diseases provided a major impetus for
the study of the distribution of sodium and potassium channels.
In unmyelinated nerves, each patch of membrane has both sodium
and potassium channels and each area is capable of generating
an action potential so that propagation of the impulse in these
fibers is said to be "continuous." Sodium channels in mammalian
myelinated nerve are localized at the node of Ranvier where
there are few potassium channels. Acute demyelination experiments with lysolecithin demonstrate no sodium channels in the
internode and the internode remains inexcitable. When myelin
breakdown is slower, as exemplified by the diphtheria toxin model
in peripheral nerve, excitability of the internodal axonal membrane

is demonstrable within 4-5 days as the fiber is demyelinating. This excitability implies the presence of sodium channels in the membrane. The mechanism of appearance of the sodium channels is unknown. Possibilities include: a) the migration of channels from the original node of Ranvier, b) the appearance de novo of sodium channels, or c) the "uncovering" or the functional activation of small numbers of already present channels.

Question 1
Are sodium channels susceptible to modulation by exogenous factors either nonspecifically or specifically (e.g., antibodies as in myasthenia gravis)? In addition to the process that causes demyelination, could additional functional impediment be related to such factors?

Following demyelination with lysolecithin, the appearance of sodium channels in paranodal or internodal regions occurs as an early sign of reorganization of the demyelinated nerve. Before the resynthesis of myelin by Schwann cells, a patchy distribution of sodium channels is found as inferred from two sources: a) electrophysiologically, viz., the spatial distribution of membrane currents in the axon, and b) morphologically using special stains. With subsequent remyelination, a new internodal distance is created that is shorter than before. It is not known whether insertion of sodium channels is the actual determinant of internodal length.

The peripheral and central nervous systems do not seem to differ in organization of sodium and potassium channels, but the response of central myelinated nerve fibers to demyelination is yet to be examined by biophysical methods.

Question 2
Is the function of the Schwann cell in the PNS (and the oligodendroglia in the CNS) solely that of providing electrical insulation?

The role of calcium in conduction in demyelinated fibers was
raised, and it was recognized that calcium could either hinder
or help conduction. Were sufficient extra calcium added to
the external medium surrounding the nerve, conduction could
be blocked since Ca raises the threshold for sodium channel
opening. Yet, in the lobster axon, injection of Ca into the
axon can relieve a block resulting from repetitive activation
of the fiber.

Question 3
What is the fate of K that has left the axon in an impulse?
Are glial cells important in buffering fluctuations of K or
could they actually take up potassium? How does potassium
get back into the cell?

In terms of physiologic consequences of injury, it is possible
that the close proximity of demyelinated fibers to each other
in the central nervous system might promote "cross talk" resulting in spontaneous activity. This could theoretically
result in clinical symptoms. "Cross talk" between axons does
occur in dystrophic mice as does "ectopic" impulse generation (8).

In diphtheria toxin induced demyelination, early dysfunction
occurs coincident with paranodal widening. This results in
increasing the capacitance and lowering the resistance of the
nerve fiber. Following these initial changes, segmental demyelination occurs. Although originally inexcitable, after
several days the demyelinated segment does acquire the ability
to exhibit continuous conduction. However, although the axon
is electrically excitable, there may still be no conduction
of an impulse because of the failure of the propagating impulse
in more normal regions of the fiber to excite the potentially
excitable demyelinated stretch of axon. The essential step in
restoring conduction is to prolong the action potential in order
to supply the extra charge needed to bring the membrane to its
firing threshold.

Question 4
In diphtheria toxin demyelination of peripheral nerve, are the
Schwann cells of unmyelinated nerve fibers also killed, and
would this affect conduction in tnose nerve fibers as well as
in the demyelinated fibers?

A central point is that in a demyelinated axon conduction is
either absent or slowed and is susceptible to block with repetitive activation. In the latter case one must assume that
there is a critical degree of paranodal demyelination and
segmental demyelination that leaves conduction on the verge of
failure, and this is very sensitive to a variety of factors.
The safety factor of transmission in demyelinated nerve may
be close to unity, and change that increases the safety factor
will favor conduction. It is known that in experimental models
of demyelination, lowering the temperature of the preparation
favors conduction by lengthening the action potential, and the
beneficial effects of lowering body temperature is well-
documented in multiple scleroses. For example, the spatial
extent of a foveal scotomata can be decreased by lowering
body temperature and increased by raising it.

In principle, given the appropriate manipulation, it is possible
to restore conduction in demyelinated nerve fiber. This can
be done by agents that prolong the action potential, as has
been achieved either by blocking potassium channels (4-amino-
pyridine) or by slowing the closing of sodium channels (scorpion venom).

Question 5
What are the consequences of long-term block of impulse transmission on synaptic transmission in the central nervous system?

Question 6
Are the acute symptoms of a patient with MS due to edema or to
actual demyelination? What is the effect of edema on conduction?

Question 7

Is the apparent symptomatic recovery between episodes actually
due to reorganization of demyelinated axons so permitting continuous conduction in the absence of myelin?

Question 8

Could an altered intracellular ionic environment (brought about
by therapeutic drugs, toxins, electrical activity, etc.) alter
the neuronal or glial susceptibility to immunologic or viral
attack?

Question 9

Is there evidence for the long-term persistence of electrically
excitability (continuous conduction) over long distances in demyelinated, large diameter axons? In other words: Is continuous conduction as described by Bostock and Sears a temporary or long-term phenomenon?

Question 10

What are the characteristics and nature of impulse transmission
in developing nerve fibers before myelination?

Question 11

How is excitability restored?

MECHANISMS OF IMMUNE INJURY (see Arnason and Waksman, this
volume)
The mechanisms by which the immune system could injure the
nervous system are multiple and relate to the various limbs
of the immune system.

A. Antibody mediated damage can occur according to the following
schemes: a) complement dependent lysis as seen with demyelination
in culture, b) antibody induced modulation of surface receptors
affecting function as has been reported for acetylcholine receptors in myasthenia gravis, c) antibodies against a specific
surface component of the cell which can stimulate the cell as

exemplified by antibody against the receptor for thyroid
stimulating hormone producing hyperthyroidism; in addition,
antibodies against caudate and subthalamic neurons have been
implicated in Sydenham's chorea, and d) antibody has been
reported to stimulate pump activity in the RBC.

Question 12
How accessible are specific tissue components of the nervous
system to antibody?

B. Immune complexes can produce vascular injury following
subendothelial deposition. The cerebrospinal fluid contains
complement components which are at significantly lower concentrations than those found in the blood; the optimum concentration for immune complexes to adhere to cells is in fact the
concentration of complement in the CSF. Immune complexes occur
in multiple sclerosis, but at low concentrations. In certain
viral infections, e.g., progressive rubella panencephalitis,
it has been possible to dissociate immune complexes and
identify rubella antigens (2).

Question 13
Are there immune complexes near lesions in MS? What are the
antigens associated with immune complexes in MS?

C. Cell mediated immunity involves damage triggered by
sensitized T cells. Antigen is presented on the surface of
antigen presenting cells (macrophages) in the context of
products encoded by the I region of the major histocompatibility
complex. Release of various soluble mediators by the T cell
recruits macrophages and ultimately other inflammatory cells.
Various subsets of T cells in the immunoregulatory network
(discussed later) participate in regulating both cellular and
humoral immune responses. Macrophages, natural killer cells,
and cells directed by antibody bound to their surface Fc receptors may also participate in nervous system injury (antibody dependent cell-mediated cytotoxicity; ADCC).

The nature of immune reactivity involves immunologic memory and specificity. If multiple sclerosis is an autoimmune disease, clearly the nature of the antigen(s) providing the specificity must be defined. Despite multiple studies, there is no conclusive evidence that MS patients show uniform immune reactivity to myelin basic protein or other white matter constituents (e.g., oligodendrocytes). Nevertheless, there is a reasonable possibility that the antigen(s) involved in an autoimmune attack in MS are located on the oligodendrocyte.

Experimental allergic encephalomyelitis has long been studied as a possible model for multiple sclerosis. Initial work with white matter preparations prior to 1950 gave way to studies in which the encephalitogenic protein of myelin (myelin basic protein) could be identified and sequenced; in the guinea pig the minimal size encephalitogenic peptide is 9 amino acids in length. A return to the study of more "crude" preparations has recently occurred as it has become clear that myelin basic protein may not be the unique sensitizing antigen in multiple sclerosis. Furthermore, it appears that more than basic protein may be required to induce chronic relapsing EAE in animals. This form of EAE resembles MS histologically and can now be produced in mice of the SJL strain. This strain is well-known as having an "autoimmune tendency."

Question 14
Is the stripping of myelin by macrophages in EAE a primary or secondary event and how does it occur?

Question 15
What is (are) the causative antigen(s) of chronic relapsing EAE? What triggers relapses in chronic EAE?

Question 16
Will sublethal CNS injury by diphtheria toxin or other toxic or viral agents lead to autoimmune disease on a "favorable" immunogenetic background?

The role of various "factors" in producing the demyelinating
lesions of EAE has been investigated. Protease inhibitors
inhibit lesion formation in EAE, while aprotinin enhances them.
Lymphokines injected into the dorsal root ganglia give inflammation but no demyelination unless anti-galactocerebroside
antibody is also injected. Using the rabbit eye as a model,
demyelination is only seen when both lymphokines and anti-
myelin antibodies are injected, implying that ADCC may be
superimposed on the cell-mediated immune response.

MECHANISMS OF VIRAL INJURY   (see Fields and Weiner, this
volume).

Viruses can enter the nervous system by a variety of routes
blood, retrograde axonal transport in nerves, and perhaps on
cellular elements that carry them past the blood-brain barrier
into the nervous system. Specificity of viruses is to a large
extent determined by the interaction of specific viral proteins
with cell surface components (receptors). Examples of systems
where cell receptors determine cell tropism include poliovirus
(anterior horn and motor cells), reovirus type 3 (neurons),
and reovirus type 1 (ependymal cells). In some instances the
nature of the receptor to which viruses bind is known (e.g,
influenza), but for most viruses, no detailed characterization
of viral receptor structures has been done.

Tropism of viruses for the oligodendrocyte (e.g., certain
coronaviruses, J-C virus in multifocal leukoencephalopathy)
may or may not be due to specific cell surface receptor interactions. J-C virus enters astrocytes, but only damages
oligodendrocytes. It is clear that specificity is not simply
determined by the binding of a virus to the surface of a cell,
as a virus may bind to a variety of cells, while only certain
cells are capable of supporting viral replication. Virus can
enter the cell in a variety of ways, some being specific, others
non-specific. One described pathway for Semliki Forest virus
involves adsorptive endocytosis with transfer into coated pits
(4). There are various mechanisms by which a virus may injure
a cell, once inside the cytoplasm. The best current example is
poliovirus, which specifically modifies the cap binding protein.

Work on specific reovirus genes is beginning to identify gene
products responsible for properties such as viral spread and
inhibition of cellular protein synthesis. Viral persistence
is an area defined best in relation to specific viruses and
not the topic of the present report, but it is clear that many
viruses change during persistence and mechanisms by which they
persist is varied. Demyelination can be induced by a virus by
multiple mechanisms which may or may not be immune mediated.
For example, studies of the demyelinating properties of the JHM
strain of murine coronaviruses demonstrate that the strain of
virus, age of animal, and immune status of the animal determine
to what extent demyelination and oligodendrocyte involvement
occurs (see Barchi et al., this volume).

Question 17
Are there shared antigens between viral components and glia/
neuronal cell surface components? Is there a "viral receptor"
on oligodendrocytes?

No virus has been established as the "etiologic agent" of
multiple sclerosis. It is possible that a virus is not involved,
that multiple viruses are involved, that a virus was involved
initially and then disappeared, or that sensitive enough probes
have not been developed to identify a virus. In the visna
model of demyelination in sheep, retrovirus is present primarily as a provirus, cannot be seen by electron microscopy,
can be visualized with nucleic acid probes, and is recoverable
only with specific co-cultivation techniques (3).

Question 18
Are certain viral genomes detectable in MS tissue if molecular
probes of sufficient sensitivity and specificity are used?

There are multiple ways in which viruses can interact with the
immune system, and subsequently modulate it. Viruses can
infect lymphocytes and macrophages or could interact with
receptors on the surface of lymphocytes and alter lymphocyte

function. For example, reovirus type 3 generates suppressor
cells. Receptors for a variety of viruses are present on the
lymphocyte surface. In addition, it is known that there are
receptors for various neurotransmitters on lymphocytes as well
(e.g., acetylcholine receptors, dopamine receptors, histamine
receptors). Of note is that histamine receptors are present
on suppressor cells. Interferon is produced by lymphocytes
and is being studied with regard to its possible role in the
etiology and pathogenesis of MS either in terms of its immunoregulatory effects or its anti-viral properties.

Question 19
Is the lymphocyte surface altered in any way during acute or
persistent viral infections?

CONSEQUENCES OF INJURY IN RELATIONSHIP TO MULTIPLE SCLEROSIS
(see McDonald, this volume)
The following "hard facts" related to multiple sclerosis will
need to be integrated into any explanation of the disease:
a) geographical distribution, b) genetic factors, c) demyelination, d) oligoclonal immunoglobulin in the spinal fluid,
e) alterations in peripheral blood T cell phenotypes, and
f) relapses and remissions.

Geography
MS has a peculiar geographic distribution that in a broad
sense identifies high prevalence areas as being furthest from
the equator. There are problems with some of the data and
clear genetic factors that affect the geographic distribution
(low incidence in orientals, areas of extremely high prevalence).

Genetic Factors
MS does not obey Mendelian genetics. Five to ten percent of
MS patients have family members with MS and there is a 40%
incidence of MS in identical twins. In addition, there is an
increased incidence of certain histocompatibility (HLA) types
or immune response genes in MS patients. Of the diseases

linked to HLA, most are of a chronic or recurrent inflammatory
nature felt to be autoimmune or viral in origin.

Demyelination

Demyelination implies injury to myelin or the oligodendrocyte
or both. There are no oligodendrocytes in an MS plaque, though
they may be present in "shadow plaques." Pathologically, MS
has a predilection for certain sites in the nervous system,
specifically surrounding ventricular surfaces; these are perivenular in distribution.

Question 20

Which comes first in MS, demyelination or the inflammatory
reaction?

Question 21

Are there changes in myelin turnover rates in MS?

Question 22

Are oligodendrocytes targets for immune injury? Could astrocytes be the target?

Immunoglobulin in the Spinal Fluid

Immunoglobulin in the spinal fluid of MS patients is elevated
and is of a restricted heterogeneity, or oligoclonal in nature,
although the antigen(s) against which the immunoglobulins are
made is (are) not known. Recent studies using isoelectric
focusing suggest that these oligoclonal bands may be "non-
specific" for the most part, representing the products of
immunocomponent B cells carried to the CNS which then mature
to plasma cells and produce antibody in a continuous fashion,
perhaps because they are removed from the inciting antigen and
immunoregulating influences (6). Thus, the relevant autoantigen
in MS may be masked by a variety of cells present in the CNS
which produce antibodies other than those involved in disease
pathogenesis. In EAE, when animals get sick, there are no
oligoclonal bands in the CSF, but bands appear after 4-5 weeks.

This may also be seen in viral infections in animals or man.
Nonetheless, in CNS infections, although there may be oligoclonal bands for a period of time, they then disappear. Thus
in MS patients something may be continually stimulating the
plasma cells in the CNS to produce antibody. In SSPE it is
possible to absorb the oligoclonal bands with measles antigen.
By contrast, all attempts to absorb the majority of oligoclonal
bands in MS with viral and brain antigens have been unsuccessful
(anti-viral antibody can be absorbed, but the major bands remain).
There are slightly increased numbers of lymphocytes in the CSF
of MS patients during an attack, and lymphocytes from MS patients
are found to be in G1, G2, or S phases of the cell cycle,
suggesting ongoing disease activity even between attacks (7).

Question 23
How and why are B cells trapped in the brain of MS patients?

Question 24
What is the lymphocyte traffic through the normal brain and
after its injury, e.g., through chronic virus infection?

Alterations in Peripheral Blood T Cell Phenotypes
In MS patients during active disease there is a decrease in
functionally active non-specific suppressor T cells (1) and a
decrease in actual number of T suppressor cells as measured by
immunofluorescence with monoclonal antibodies. Loss of suppressor T cells as measured by monoclonal antibodies is not
seen in Guillain-Barre or acute disseminated encephalomyelitis
(9). There is also a decrease in cells bearing the Fc receptor
for IgG during attacks (5). These changes return to normal
when the acute attack is over, are not seen in a variety of
other nervous system diseases, and are unexplained. These
changes are dramatic and represent the disappearance from the
blood of as much as 20% of the T cell population. Changes in
immunoregulatory T cells are also seen in lupus erythematosus,
juvenile rheumatoid arthritis, graft vs host disease, and

hypogammaglobulinemia. In addition, lymphocytotoxic antibodies
are present in the sera of MS patients and of patients with
lupus erythematosus.

Question 25
Are there antigenic determinants shared by oligodendrocytes and
suppressor T cells?

Question 26
Can Ag-specific suppressor cells be identified in MS?

Question 27
What are suppressor cell kinetics (blood and CSF) in acute and
chronic infections, EAE, and MS?

Question 28
Where do suppressor cells go during MS attacks?

Question 29
Against which lymphocyte subpopulations do lymphocytotoxic
antibodies in MS react? Do they react with CNS elements?

Question 30
What are the immune abnormalities in the Guillain-Barre syndrome, especially in cases with recurrent Guillain-Barre?

Relapses and Remissions
Relapses and remissions remain unexplained. Possible explanations (not mutually exclusive) include: a) physiologic changes
in conduction along demyelinated nerve secondary to changes
in ionic environment, edema, or extracellular factors (anitbody,
toxins); b) recurrent viral infections or alterations in an
ongoing persistent viral infection; c) changes in immunoregulatory cells which themselves would need to be "triggered"; or
d) MS may be ongoing "all the time" with exacerbations of an
ongoing process.

CONCLUSION

A major theme to emerge from the group discussion was the need to develop an investigative approach that begins at the cellular level of neural-glial interactions and to define in a mechanistic way the nature and type of injury that can occur. This requires identification of surface structures on target cells, of the "specific" antigen(s) that may be involved in an autoimmune attack, viral gene products responsible for injury or for the modulation of the immune system, and the nature of injury by toxins. A primary difficulty in approaching multiple sclerosis is the difficulty in establishing what the early events actually are and where they occur in the CNS.

Question 31

What are the details of injury to oligodendroglia due to several types of injurious agents and mechanisms, i.e., toxins, viruses, immune mediated? (The details should involve specific morphologic and functional analysis of plasma membrane, the cell cytoskeleton, the cellular metabolic machinery.)

Question 32

Is it possible to measure the rate of myelin synthesis in oligodendroglia after gentle demyelination with a detergent such as lysolecithin?

Question 33

What is the "earliest" pathologic change in organelles seen in oligodendrocytes surrounding active demyelination in MS and what are analogous changes in EAE, virus infections, or toxic injury?

REFERENCES

(1) Antel, J.P.; Arnason, B.G.W.; and Medof, M.E. 1978 Suppressor cell function in multiple sclerosis: Correlation with clinical disease activity. Ann. Neurol 5: 338-342.

(2) Coyle, P., and Wolinsky, J.S. Characterization of immune complexes in progressive rubella panencephalitis. Ann. Neurology, in press.

(3) Haase, A.T.; Stowring, L.; Narayan, O.; Griffin, G.; and Price, D. 1977. Slow persistent infection caused by visna virus: role of host restriction. Science 195: 175-177.

(4) Helenius, A., and Simons, K. 1980. Mechanisms of animal virus entry into cells. In The Molecular Basis of Microbial Pathogenicity, eds. H. Smith, J.J. Skehel, and M.J. Turner, pp. 41-54. Dahlem Konferenzen. Weinheim: Verlag Chemie.

(5) Huddlestone, J.R., and Oldstone, M.B.A. 1979. T suppressor (T$_G$) lymphocytes fluctuate in parallel with changes in the clinical course of patients with multiple sclerosis. J. Immunol. 123: 1615-1618.

(6) Mattson, D.H.; Roos, R.P.; and Arnason, B.G.W. 1980. Isoelectric focusing of IgG eluted from multiple sclerosis and subacute sclerosing panencephalitis brains. Nature 287: 335-337.

(7) Noronha, A.; Richman, D.P.; and Arnason, B.G.W. 1980. Detection of in vivo stimulated cerebrospinal fluid lymphocytes by flow cytometry in patients with multiple sclerosis. New Eng. J. Med. 303: 713-717.

(8) Rasminsky, M. 1978. Ectopic generation of impulses and cross-talk in spinal nerve roots of "dystrophic" mouse. Ann. Neurol. 3: 351-357.

(9) Reinherz, E.L.; Weiner, H.L.; Hauser, S.L.; Cohen, J.A.; Distaso, J.A.; and Schlossman, S.F. 1980. Loss of suppressor T cells in active multiple sclerosis: analysis with monoclonal antibodies. New Eng. J. Med. 303: 125-129.

Neuronal-glial Cell Interrelationships, ed. T.A. Sears, pp. 287-302.
Dahlem Konferenzen 1982. Berlin, Heidelberg, New York: Springer-Verlag.

Recovery and Function After Demyelination

H. Bostock and W. I. McDonald
Sobell Dept. of Neurophysiology and Dept. of Clinical Neurology
Institute of Neurology, London WC1N 3BG, England

Abstract. Recovery even from a severe acute neurological deficit is usual in the early stages of multiple sclerosis (MS).
The development of the deficit can be accounted for fairly
readily by the known physiological abnormalities produced by
demyelination, but the mechanisms of recovery from it are poorly
understood. Remyelination can restore conduction in the CNS,
but few fibers in the lesions of MS show evidence of new myelin
sheaths. Continuous conduction can develop in demyelinated
peripheral fibers, and there is some evidence for conduction
in persistently demyelinated fibers in MS. It is unknown whether the recovery mechanisms which compensate for degenerative
central lesions contribute to recovery from demyelination. Recent experimental work suggests that it might be possible to
improve conduction in demyelinated central nerve fibers by prolonging the action potential pharmacologically.

INTRODUCTION

One of the striking clinical features of MS is the extent of

recovery of function which can occur after an acute relapse.

For example, some 90% of patients with acute optic neuritis

(one of the most common manifestations of MS) regain normal

vision in a matter of a few weeks even when the acuity is re-

duced below the level at which it is possible to read the larg-

est letters on the standard charts. However, although there is

clinical recovery, the lesion persists: in 90% of patients an

abnormality in the visual evoked potential (VEP) remains (20).

Similar observations have been made in relation to episodes in-

volving other sensory pathways, and it is common at postmortem

to find plaques in tracts, the function of which was judged to
be clinically normal during life.

The central problem in understanding the mechanism of remission
in MS is thus that of explaining the restoration of function
despite the persistence of the lesion. Before considering the
processes underlying recovery, it is necessary to review the
mechanisms of functional loss.

THE PHYSIOLOGICAL CONSEQUENCES OF DEMYELINATION
Experimental work over the past 20 years has established that
demyelination of central and peripheral nerve fibers interferes
with conduction of nerve impulses in a number of ways (9,17).
Broadly speaking, 'large' demyelinating lesions are associated
with complete conduction block, while smaller lesions permit conduction to continue in a proportion of fibers. But conduction
is slow and the refractory period of transmission of the fibers
is prolonged, resulting in impairment of their ability to transmit long trains of impulses at physiological frequencies. If
such fibers are continuously stimulated above a critical frequency, the refractory period increases until conduction is completely, though reversibly, blocked. Finally, it has recently
been shown that a demyelinated portion of a nerve fiber is ex-
quisitely mechano-sensitive and that impulses can arise spon-
taneously in the demyelinated region (12,34).

Different fibers traversing the same lesion may show different
degrees of conduction abnormality, in keeping with the variety
of morphological changes which are found. Some fibers may lose
one or more complete internodal segments, while others may show
shorter lengths of demyelination ranging from nodal widening
by a few μm to several hundred μm of partial- or full-thickness
loss of the myelin sheath. Consecutive internodes on a single
fiber often show different changes.

While there is physiological evidence for the broad relationships
between the extent of demyelination and the various forms of

conduction defect (14,26) more precise correlation awaits the
development of techniques permitting the physiological and morphological study of the same identified fiber.

Clinical Implications

The application of evoked potential techniques to patients with
MS has provided evidence for the existence in man of the same
types of conduction defect as those seen in experimental animals
(9). It is, however, only possible at present to make fairly
crude interpretations of the clinical features of MS in the light
of the known conduction defects.

Severe weakness or sensory loss can be accounted for readily by
complete conduction block in a large proportion of the fibers
subserving those functions. The detailed interpretation of the
relationship between less complete lesions and symptomatology
is more difficult partly because of the difficulty of determining
physiologically the proportion of fibers blocked, and partly because of the relative insensitivity of many clinical tests.

Unequal slowing of conduction in different fibers passing through
a lesion would be expected to impair those functions - such as
vibration sensibility - which depend on the delivery of precisely
timed bursts of impulses. Vibration sense is usually impaired
in multiple sclerosis.

It is likely that impairment of the ability of nerve fibers to
transmit trains of impulses at physiological frequencies plays
an important part in muscle weakness. Increasing refractory
period is plausible as a contributing factor to the progressive
increase in weakness observed during sustained exercise in patients with demyelinating lesions affecting the motor pathways.

The abnormal mechano-sensitivity of demyelinated nerve fibers
provides a satisfactory explanation of the very characteristic
symptom of tingling on neck flexion, and the long sustained,
resting spontaneous discharges could similarly account for the
continuous tingling so common during acute relaspses of MS.

Occasionally patients with MS experience short-lived episodes
lasting only a matter of a few minutes. Some of these, such as
the transient worsening produced by small elevations of body
temperature (25), are explicable on the basis of the known properties of demyelinated nerve fibers. Others are less readily
explained and it has been suggested that there might be circulating factors which interfere with either synaptic transmission or conduction in the nerve fibers themselves. The evidence
is conflicting both in experimental allergic encephalomyelitis
and in MS. Complement-dependent serum factors have been reported
to interfere with 'complex bioelectric activity' in vitro in
the mouse neocortex culture preparation and in the isolated per-
fused frog spinal cord, but convincing evidence that the effects
are specific is lacking (38). Peripheral demyelination can be
produced by intraneural injection of experimental encephalomyelitis serum (39), but it is not known what contribution the serum
factors make to the symptoms of the disease.

Summary

Although the correlation between clinical abnormalities in MS and
the physiological properties of demyelinated nerve fibers is possible only in rather general terms, it is clear that much of the
symptomatology of the disease is attributable to the effects of
the lesion on conduction through it.

RECOVERY OF FUNCTION

The usual course of events, at least in the earlier episodes of
MS, is that the symptoms develop over a matter of a few days or
a week or two, persist for several weeks, and gradually subside
over a month or two. Variations are, however, frequent. It is
not uncommon for the initial stage of improvement to be quite
rapid: in optic neuritis, for example, visual acuity may improve
from the ability just to count figures to near normal (6/9) in
less than a week. On the other hand, after a devastating relapse,
improvement may continue for more than a year.

In elucidating the mechanisms of recovery, it is necessary first
to review what is possible in the way of reversing the morphological and physiological abnormalities and second to consider whether
the time course of such changes corresponds with the time course

of clinical recovery. Recovery can involve remyelination of
the affected fibers, but there is also good evidence that conduction can be restored without remyelination. Even with persisting conduction block, adaptive changes may occur in the surviving parts of the nervous system.

Remyelination
It has been known for some 20 years that under certain circumstances remyelination can occur in the central nervous system
(4). New internodes are formed but appear to remain abnormally
thin and short for at least a year (8,11). There appears to
be some relationship between axon diameter and internodal length
in the remyelinated fibers, but chains of short internodes - some
of which may be very short indeed (5-15 μm) - occur.

Recent experiments on the lesion induced by the direct microinjection of lysophosphatidyl choline (LPC) into the spinal cord
of the cat have established that remyelination by oligodendrocytes is capable of restoring conduction in previously blocked
fibers (32). The earliest evidence of transmission through the
lesion appears at about 14 days, the time at which remyelination
is first seen electron microscopically. The proportion of blocked
fibers diminishes rapidly over the following month. The refractory period of transmission, which gives a guide to the ability
of the fibers to transmit trains of impulses, is at first much
prolonged (about 4 times normal) but gradually returns to normal
after about 3 months, a time at which the myelin sheaths are still
abnormally thin and short.

What part does restoration of conduction of remyelination play in
recovery from the relapse of multiple sclerosis? Probably relatively little. Although the time course would be appropriate,
the amount of remyelination in MS is very limited. There is,
however, clinical evidence that some form of conduction can occur
in the absence of remyelination (37):

Case report. A 40 year old woman with established multiple
sclerosis was admitted to hospital 4 days before she died from
a massive pulmonary embolus. Visual acuity was recorded by two
separate examiners on admission. The patient was able to count
fingers with each eye and to read with telescopes. At necropsy
both optic nerves were exhaustively examined by light and electron microscopy. Complete transverse sections were taken at many
levels through the whole length of the nerves. Many areas in
each nerve were examined by electron microscopy and at one level
in one nerve a montage of a complete transverse section of the
nerve was constructed from electron micrographs. Many intact
axons were present in each nerve. Myelin sheaths were, however,
completely absent in one nerve. In the other, myelin was present only over a length of 3mm in a small group of fibers at one
edge of the nerve. The total length of demyelinated nerve was
30mm on one side and 27mm on the other. Since this patient had
some preserved visual function, it is clear that the demyelinated
axons were conducting. We have no other information about conduction in this patient since she died before evoked potential
techniques became routine in clinical practice.

Recovery of Conduction Without Remyelination
In this section we describe experimental results on peripheral
nerve which indicate how demyelinated fibers can conduct in the
absence of remyelination, and we consider the evidence that
similar mechanisms might operate in MS.

Continuous Conduction
Recordings from experimentally demyelinated peripheral nerves
have sometimes been interpreted in terms of continuous conduction along lengths of demyelinated axon. The evidence has
usually been ambiguous, but Lehmann & Ule found both complete
breakdown of myelin in the region of a granuloma induced in
rabbit saphenous nerve and very slow conduction across it (15).
More detailed physiological studies at first produced contrary
evidence. McDonald & Sears produced a focal demyelinating lesion in cat dorsal columns with diphtheria toxin and found that

when the lesion was large conduction in single fibers was blocked
within the lesion, although it persisted on either side (21,22).
With a smaller lesion, conduction persisted in many fibers but was
slow and refractory. To determine the mechanism of conduction in
such cases, Rasminsky & Sears devised an elegant technique for
recording from undissected single fibers in rat ventral roots (26).
With a pair of electrodes 400 μm apart, they recorded the longitudinal currents flowing in the root, and therefore, in the single
active fiber within it, at a series of positions along the root.
They found that the usual saltatory pattern of conduction, with
current entering and leaving the fiber almost exclusively at the
nodes, was preserved in diphtheritic demyelination, but that internodal conduction times could be prolonged as much as 25-fold before conduction was blocked. A parallel theoretical study suggested that conduction block would be the inevitable consequence
of complete segmental demyelination (14). It looked as if previous
presumptions of continuous conduction in clinical and experimental
demyelination must have been mistaken, but (a) Rasminsky & Sears
had recorded selectively from large fibers in a lesion known to
produce predominantly paranodal, rather than segmental demyelination in large fibers (13); and (b) the theoretical model was made
without knowledge of either the active or passive electrical properties of the internodal axon membrane.

An improved technique for recording longitudinal currents later
demonstrated clearly that continuous conduction can occur in
peripheral fibers demyelinated with diphtheria toxin (2). Using
electrodes only 120 μm apart, internodes were occasionally found
in which strong inward (sodium) currents were generated over each
100 μm that the electrodes were displaced, in contrast to other
parts of the fiber where inward currents were confined to the nodes
as in normal fibers. The continuous conduction, most readily
seen in small fibers at reduced temperatures, occurred at velocities
of 1-2m/s, or about 1/20th-1/40th of the velocities expected for
normal stretches of the same fibers. Supporting evidence that
this type of conduction was occurring along myelin-free segments
of nerve has come from (a) estimates of membrane capacity from

the foot of the continuously conducted action potential (2),
and (b) the complete absence of remyelination in skip-serial
electron microscope sections through a 6-day lesion in which
a continuously conduction fiber had been recorded (Pullen &
Bostock, unpublished observations).

Discontinuous Conduction
Conduction in fully demyelinated axons is not always continuous.
Nerve fibers demyelinated with lysolecithin are initially
blocked but start to conduct impulses prior to remyelination
(33), and longitudinal currents have now been recorded in this
lesion (1). At 6 days, conduction across a demyelinated segment appeared saltatory, although slow, with several new foci
of inward membrane current resembling nodes. Remyelination was
completely absent at this stage, although due to commence a
day or two later. It seems, therefore, that the concentration
of sodium channels at nodes starts prior to remyelination. Conduction velocities increased dramatically accompanying remyelination, but before remyelination the discontinuous conduction in the lysolecithin lesion was about as slow as the continuous conduction in the diphtheritic lesion.

Relevance to Multiple Sclerosis
What is the evidence that the phenomena observed in demyelinated
rat spinal roots are relevant to central demyelinating lesions
and, in particular, to those of MS? There is as yet no experimental evidence bearing on the possibility of continuous conduction developing in demyelinated central axons. However, all
the physiological properties of normal, degenerating, demyelinated, and remyelinated peripheral fibers have, when looked for,
been found in central nerve fibers in corresponding states (17).
It therefore seems likely on physiological and morphological
grounds that the phenomena of conduction without remyelination
have their counterpart in the central nervous system of man.
Conduction in persistently demyelinated fibers could account
not only for the persistence of limited clinical function as
in the patient described above, but it could also provide a

reasonable explanation of the delays characteristic of cortical
evoked responses in MS.

Abnormally long latencies have been observed in MS patients for
somatosensory evoked responses recorded from the spinal cord and
from the cortex, for early components of the auditory evoked
response ('brain stem evoked responses'), and for the cortical
responses evoked by flash or by reversing a checkerboard visual
stimulus (20,24). To summarize briefly the results obtained
with pattern reversal (10): (a) in the acute phase of optic
neuritis, loss of visual acuity is accompanied by decreased amplitude and increased latency of the VEP; (b) during the recovery
phase, the amplitude of the VEP increases as the visual acuity
improves, but the latency remains stable, increased by an average
of 30-40ms; (c) a further attack could produce an additional
delay; (d) a high proportion of MS patients without visual symptoms have 'silent' lesions, indicated by abnormal latency VEPs.
These observations are most simply explained if demyelinated
stretches of optic nerve can, after a recovery period, conduct
abnormally slowly, An estimate of the conduction velocity change
necessary to account for the delays observed can be obtained
by comparing the VEP delays (up to 100ms, mean approximately 35ms)
with the lengths of plaques of demyelination found in MS patients
at post mortem of 3-30mm, mean 10.5 mm (19). The ranges and
means fit well if the fibers mediating this response conduct at
about 0.3m/s. Comparable figures are not available for plaques
in the somatosensory and auditory pathways. The relatively much
smaller delays found in auditory brain stem evoked responses (28)
presumably reflect in part the fact that, unlike the optic nerve,
the cochlear nerve is not part of the CNS, so that the conduc-
tive path subject to demyelination is shorter. Remarkably slow
as a velocity of 0.3m/s may appear, it is not unreasonable in
view of what is known of continuous conduction in demyelination.
The range of conduction velocities in human optic nerves has not
been determined, but given that 92% of the fibers are about 1 μm
in diameter, and assuming a Hursh factor of 6, a typical value
of about 6m/s is likely. On this reckoning, 0.3m/s is 1/20th of

normal and compares with the estimates for continuous conduction of 1/20th to 1/40th of normal (2).

The simple interpretation of long latency evoked responses in
terms of altered axonal conduction alone has been doubted, and
both retinal (23) and cortical (6) delay sites have been suggested. At present, however, the only mechanisms known that
can account for the observed variation in the delays in terms
of the known pathology of the disease is continuous conduction.

Implication for Symptomatic Treatment

The demonstration that demyelinated axons become electrically
excitable following demyelination and can transmit impulses in
favorable circumstances has important implications for the treatment of MS patients. It suggests that the conduction block that
is the more usual consequence of demyelination is due to the impedance mismatch at the junction between myelinated and demyelinated parts of the axon, rather than to the inability of the demyelinated parts to conduct. The block occurs because the driving
potential at the last active node, conducted via the axial resistance of the last internode, does not last long enough to
depolarize to threshold the demyelinated segment with its large
input capacity. Bostock & Sears argued that because the internodal axial resistance should scale inversely with fiber diameter,
whereas the input impedance of a demyelinated segment should
scale inversely with the 3/2 power of the diameter, the impedance
mismatch should be less for smaller fibers (2). Transmission
across a fully demyelinated plaque should therefore occur more
readily for fibers of the optic nerve (mostly about 1 μm diameter) than, for example, for the large dorsal column fibers
studied by McDonald & Sears (22).

The crucial point as far as treatment is concerned is that any
factor which increases the safety factor for nervous transmission should enable more demyelinated fibers to conduct. The
case for attempting a pharmacological modification of safety
factor was made by Schauf & Davis (29), even though they supposed

continuous conduction was impossible, and the case is now much
stronger (30). Schauf & Davis pointed out that the safety factor (defined by Tasaki as the ratio between the current available and the minimum required to excite the next segment) could
be increased in two ways: by prolonging the action potential
or by reducing the threshold for excitation. There is ample
evidence that prolongation of the action potential by cooling
improves conduction in MS patients (9), and treatment designed
to reduce nerve thresholds by reducing serum ionized calcium
has also produced transient improvements (7). Of the two approaches, action potential prolongation is the more promising,
since any reduction in thresholds would be expected to exacer-
bate the 'positive' symptoms in MS due to ectopic excitation.
According to the Hodgkin-Huxley model of nerve excitation, action potential prolongation can be achieved by delaying sodium
inactivation or by blocking or delaying potassium currents, as
well as by reducing temperature. Scorpion venom delays sodium
inactivation and helps overcome conduction failure in experimentally demyelinated fibers (3) but could not be countenanced
for clinical trial in MS. 4-Aminopyridine blocks mammalian potassium channels and can help overcome conduction failure in
experimental demyelination, although it has no effect on normal
nodes (31). Since this substance has already been used clinically (e.g., in myasthenia gravis, (16)), a preliminary clinical
trial on MS patients has been initiated. The initial results
have, however, been rather discouraging. Intolerable side effects developed before there was any clinical improvement.

Mechanisms of Recovery
One recovery process required for continuous conduction is the
spread or generation of sodium channels along the axonal membrane previously covered by myelin. The normal absence of
sodium channels under the myelin was first suggested by the
TTX-binding experiment of Ritchie & Rogart (27) and has recently
been confirmed by Chiu & Ritchie (5), who have caused acute
paranodal and internodal demyelination of voltage-clamped fibers
and observed no increase in the sodium currents. Following

experimental demyelination either with diphtheria toxin (2) or
lysolecithin (1), electrophysiological evidence for internodal
sodium currents had been found as early as 4 days, and conduction through the lesion (and without remyelination) as early
as 6 days. It may be significant that this time scale is similar to that already mentioned for the rapid initial recovery
sometimes seen in optic neuritis. A much slower improvement
is, however, more common, and other recovery processes not
peculiar to continuous conduction, are likely to be involved.

It is possible that other factors may influence the function of
nerve fibers in a plaque of MS, but proof of their involvement
is lacking. Edema is a prominent histological feature of the
acute lesion, and it is possible that swelling in regions where
there is little room for expansion - e.g., the scleral and optic
canals - might produce pressure block in critically damaged
fibers. Such a mechanism would be consistent with the rapid
improvement often induced by treatment with steroids. It is
also conceivable that products of myelin and cellular breakdown
in the plaque might interfere with conduction but this possibility
has not been examined experimentally.

Adaptation in Surviving Parts of the Nervous System
It is well-known that some recovery of function is possible following partial degenerative lesions (such as hemi-section of the
spinal cord or capsular hemiplegia in man) in which the question
of restoration of function in the damaged fibers themselves
does not arise. It is clear that the surviving parts of the
nervous system assume some of the functions of the damaged parts.
Little is known of the mechanisms underlying such recovery (18,36).
There is evidence for the expansion of functional connections of
surviving undamaged fibers after partial section of the posterior
columns, but whether this is due to preterminal sprouting (which
is known to occur in some parts of the central nervous system)
or to enhanced effectiveness of transmission through preexisting
synapses is unknown (36). It is conceivable that similar mechansims might operate in the presence of prolonged conduction block

induced by demyelination but the possibility has not been examined experimentally.

REFERENCES

(1) Bostock, H.; Hall, S.M.; and Smith, K.J. 1980. Demyelinated axons can form 'nodes' prior to remyelination. J. Physiol. 308: 21P-23P.

(2)* Bostock, H., and Sears, T.A. 1978. The internodal axon membrane: electrical excitability and continuous conduction in segmental demyelination. J. Physiol. 280: 273-301.

(3) Bostock, H.; Sherratt, R.M.; and Sears, T.A. 1978. Overcoming conduction failure in demyelinated nerve fibers by prolonging action potentials. Nature 274: 385-387.

(4) Bunge, R.P. 1968. Glial cells and the central myelin sheath. Physiological Reviews 48: 197-251.

(5) Chiu, S.Y., and Ritchie, J.M. 1980. Potassium channels in nodal and internodal axonal membrane of mammalian myelinated fibres. Nature 284: 170-171.

(6) Cook, J.H., and Arden, G.B. 1977. Unilateral retrobulbar neuritis: a comparison of evoked potentials and psychophysical measurements. In Visual Evoked Potentials in Man: New Development, ed. J.E. Desmedt, pp. 450-457. Oxford: Clarendon Press.

(7) Davis, F.A.; Becker, F.O.; Michael, J.A.; and Sorensen, E. 1970. Effect of intravenous sodium bicarbonate disodium edetate (Na$_2$ EDTA), and hyperventilation on visual and oculomotor signs in multiple sclerosis. J. Neurol. Neurosurg. Psychiat. 33: 723-732.

(8) Gledhill, R.G., and McDonald, W.I. 1976. Morphological characteristics of central demyelination and remyelination: A single-fibre study. Annals of Neurology 1: 552-560.

(9)* Halliday, A.M., and McDonald, W.I. 1977. Pathophysiology of demyelinating disease. Brit. Med. Bull. 33: 21-27.

(10)* Halliday, A.M.; McDonald, W.I.; and Mushin, J. 1977. Visual evoked potentials in patients with demyelinating disease. In Visual Evoked Potentials in Man, ed. J.E. Desmedt, pp. 438-449. Oxford: Clarendon Press.

(11) Harrison, B.M., and McDonald, W.I. 1977. Remyelination after transient experimental compression of the spinal cord. Annals of Neurology 1: 542-551.

300   H. Bostock and W.I. McDonald

(12) Howe, J.F.; Loeser, J.D.; and Calvin, W.H. 1977. Mecha-
      nised-sensitivity of dorsal root ganglia and chronically
      injured axons: A physiological basis for the radicular
      pain of nerve root compression. Pain 3: 25-41.

(13) Jacobs, J.M. 1967. Experimental diphtheritic neuropathy
      in the rat. Brit. J. Exp. Pathol. 48: 204-216.

(14) Koles, Z.J., and Rasminsky, M. 1972. A computer simula-
      tion of conduction in demyelinated nerve fibers. J. Physiol.
      227: 351-364.

(15) Lehmann, H.J., and Ule, G. 1964. Electrophysiological
      findings and structural changes in circumscript inflamma-
      tion of peripheral nerves. Progress in Brain Res. 6: 169-
      173.

(16) Lundh, H.; Nilsson, O.; and Rosen, I.J. 1979. Effects of
      4-aminopyridine in myasthenia gravis. J. Neurol. Neurosurg.
      Psychiat. 42: 171-175.

(17) McDonald, W.I. 1974. Pathophysiology in multiple sclero-
      sis. Brain 97: 179-196.

(18) McDonald, W.I. 1975. Mechanisms of functional loss and
      recovery in spinal cord damage. In Outcome of Severe Dam-
      age to the Central Nervous System, ed. R. Porter and D.W.
      Fitzsimons, pp. 23-29. Amsterdam: Elsevier.

(19) McDonald, W.I. 1977. Pathophysiology of conduction in
      central nerve fibres. In Visual Evoked Potential in Man,
      ed. J.E Desmedt, pp. 427-437. Oxford: Clarendon Press.

(20)* McDonald, W.I., and Halliday, A.M. 1977. Diagnosis and
      classification of multiple sclerosis. Brit. Med. Bull.
      33: 4-9.

(21) McDonald, W.I., and Sears, T.A. 1970. Focal experimental
      demyelination in the central nervous system. Brain 93:
      575-582.

(22)* McDonald, W.I., and Sears, T.A. 1970. The effects of
      experimental demyelination on conduction in the central
      nervous system. Brain 93: 583-598.

(23) Milner, B.A.; Regan, D.; and Heron, J.R. 1974. Differen-
      tial diagnosis of multiple sclerosis by visual evoked
      potential recording. Brain 97: 755-772.

(24) Namerow, M.S. 1978. Evoked potentials in demyelinating
      disease. In Physiology and Pathobiology of Axons, ed.
      S.G. Waxman, pp. 421-429. New York: Raven Press.

(25) Rasminsky, M. 1973. The effects of temperature on con-
      duction in demyelinated single nerve fibers. Archives
      of Neurology (Chicago) 28: 287.

(26)   Rasminsky, M., and Sears, T.A.   1972.   Internodal conduc-
       tion in undissected demyelinated fibres.   J. Physiol.
       227: 323-350.

(27)   Ritchie, J.M., and Rogart, R.B.   1977.   The density of
       sodium channels in mammalian myelinated nerve fibres
       and the nature of the axonal membrane under the myelin
       sheath.   Proc. Natl. Acad. Sci. 74: 211-215.

(28)   Robinson, K., and Rudge, P.   1977.   Abnormalities of the
       auditory evoked potentials in patients with multiple
       sclerosis.   Brain 100: 19-40.

(29)* Schauf, C.L., and Davis, F.A.   1974.   Impulse conduction
       in multiple sclerosis: A theoretical basis for modifica-
       tion by temperature and pharmacological agents.   J. Neurol.
       Neurosurg. Psychiat. 37: 152-161.

(30)* Sears, T.A.; Bostock, H.; and Sherratt, R.M.   1978.   The
       pathophysiology of demyelination and its implication for
       the symptomatic treatment of multiple sclerosis.   Neurology
       28: 21-26.

(31)   Sherratt, R.M.; Bostock, H.; and Sears, T.A.   1980.   Ef-
       fects of 4-aminopyridine on normal and demyelinated mam-
       malian nerve fibres.   Nature 283: 570-572.

(32)* Smith, K.J.; Blakemore, W.F.; and McDonald, W.I.   1979.
       Central remyelination restores secure conduction.   Nature
       280: 395-396.

(33)   Smith, K.J., and Hall, S.M.   1980.   Nerve conduction during
       peripheral demyelination and remyelination.   J. Neurol Sci.
       48: 201-219.

(34)   Smith, K.J., and McDonald, W.I.   1980.   Spontaneous and
       mechanically evoked activity due to central demyelinating
       lesion.   Nature 286: 154-155.

(35)   Wall, P.D.   1975.   Signs of plasticity and reconnection in
       spinal cord damage.   In Outcome of Severe Damage to the
       Central Nervous System, eds. R. Porter and D.W. Fitzsimons,
       pp. 35-54.   Amsterdam: Elsevier.

(36)   Wisniewski, H.; Oppenheimer, D.; and McDonald, W.I.   1976.
       Relation between myelination and function in M.S. and
       E.A.E.   J. Neuropath. & Experimental Neurol. 35: 327.

(38)   Seil, F.J.   1977.   Tissue culture studies of demyelinating
       disease: A critical review.   Ann. Neurol. 2: 345-355.

(39)   Saida, T.; Saida, K.; Brown, M.J.; and Silberberg, D.H.
       1979.   Peripheral nerve demyelination induced by intra-
       neural injection of experimental encephalomyelitis serum.
       J. Neuropath. Exp. Neurol. 38: 498-518.

* Short reading list

Neuronal-glial Cell Interrelationships, ed. T.A. Sears, pp. 303-320.
Dahlem Konferenzen 1982. Berlin, Heidelberg, New York: Springer-Verlag.

Chemical Factors Influencing Neuronal Development

L. A. Greene
Dept. of Pharmacology, New York University School of Medicine
New York, NY 10016, USA

Abstract. A number of issues are raised and reviewed regarding chemical factors that may influence neuronal development.
These issues include the types of actions that can be possessed
by chemical factors, targets, the species of factors that are
presently known, the chemical properties, and possible routes
of delivery of the factors, factor specificity with regard to
targets, temporal limitations of factor action on neurons,
cross-regulation of factor action, tissue sources of factors,
and mechanisms of factor action. Chemical factors affecting
glial development are also briefly reviewed.

INTRODUCTION

The purpose of this background paper is to raise issues concerning the known chemical factors that may influence development of neurons. Factors that influence glial development will
also be briefly touched upon. The issues will be presented in
the form of general questions with brief discussions and inclu-
sion of pertinent references.

What Aspects of Neuronal Development Can Be Influenced?

Neuronal development can be described in terms of the following
events (though not necessarily in the following order): specification of neuroblasts, neuroblast proliferation, entrance

into the post-mitotic state, migration, association into ganglia
or nuclei, differentiation (i.e., acquisition and expression of
a complex set of general and specific neuronal phenotypes such
as excitability, morphology, and capacity for chemical transmission), guidance of neuritic outgrowth, receipt and formation
of specific synaptic contacts, and interaction with glial cells.
While part of the capacity for undergoing the above events may
be intrinsic to a given neuroblast, there is abundant evidence
that a number of these developmental steps can be affected by
a variety of extrinsically supplied chemical factors. Such
factors provide means for flexibility, fine tuning, feedback
(cell-cell and cell-environment) and temporal regulation during development.

What General Classes of Influences Can Be Described?
The actions of chemical factors influencing development may be
subsumed in three general categories (see (51)). Permissive
actions are those that would not specify or regulate a given
phenotypic property but merely allow it to occur. Instances
of this would be promotion of survival or maintenance of a
general cellular capacity such as protein synthesis. Instructive actions are those that qualitatively alter the phenotypic
properties of a target cell. Examples would be promotion of
neurite outgrowth from an otherwise undifferentiated cell or
the induction of synthesis of a particular neurotransmitter-
related enzyme. Modulatory actions are those that quantita-
tively (but not qualitatively) alter phenotypic properties.
Example would be to regulate the amount (but not type) of
neurotransmitter synthesized or to alter the length (but not
presence) of neurites.

What Chemical Factors Have Been Uncovered?
Chemical factors influencing neuronal development can be put
into several categories.

Hormones. Clinical and experimental observations have indicated that thyroid hormone profoundly affects the number as well as degree of maturation of CNS neurons (for review see (3)). The overall effect of thyroid hormone appears to be to convert proliferating neuroblasts to the post-mitotic state and to "guide" them into the differentiated state. Whether this hormone would be permissive or instructive with respect to differentiation is not clear. It has been suggested that thyroid hormone could function as a "clock" for phasing and regulating the period of neuroblast proliferation (20).

A number of behavioral and biochemical studies have indicated that the developing brain is also a target for sex hormones and that fetal or neonatal exposure to particular hormones can profoundly influence adult sexual and social behavioral patterns (for review see (40)). The physiologic basis for such effects are not clear, but one in vivo study (42) has demonstrated that the morphology and synaptic patterns of specific neurons in the preoptic area are influenced by androgens. This suggests that such hormones may have modulatory or instructive effects on their neuronal targets.

Corticosteroids appear to have a variety of possible effects in the peripheral nervous system. In both in vivo and in vitro experiments, corticosteroids have been found to greatly increase the number of observable small intensely fluorescent (SIF) or chromaffin-like cells in sympathetic ganglia (15). It is not clear whether this effect is due to increased proliferation, enhanced survival, or alteration of differentiation. Another well-described, apparently instructive action of corticosteroids is to regulate and maintain phenylethanolamine-N-methyltransferase levels in adrenal medullary cells (55). Corticosteroids have also been reported to antagonize the ability of nerve growth factor (NGF) to promote neurite outgrowth from cultured adrenal medullary cells and sympathetic neurons (50), as well as to synergize NGF's capacity to increase tyrosine hydroxylase

activity in cultured sympathetic ganglia (38). Finally,
corticosteroids have been shown to influence the choice of
neurotransmitter type synthesized by sympathetic ganglia in
culture; hormone treatment significantly increases tyrosine
hydroxylase activity and greatly suppresses choline acetyltrans-
ferase activity (36). Corticosteroids have also been found to
regulate specific enzymatic activities in cultured CNS tissue
and there are regionally specific binding sites for such hormones in the brain (see (34,40)).

Neurotransmitters. Neurotransmitters have been postulated to
play direct nonsynaptic roles in development (see (27) for
review). This possibility is suggested by the observations
that low, but significant, levels of several neurotransmitters are present in the developing CNS well before synaptogenesis.

Ions. The ionic environment has been found to influence neuronal
development in vitro. Elevated K^+ has been reported to enhance neuronal survival in cerebellar cultures (28) and to
increase the number of morphologically identifiable neurons
present in cultured dorsal root ganglia (DRG) (48). The latter
effect appears to be due to the enhancement of morphological
differentiation of cells not initially recognizable as neurons
(8). In addition, depolarizing agents were reported to bring
about nuclear division of adult motor neurons in culture (10).
While the above types of culture conditions could mimic otherwise synaptically mediated effects, they could also be representative of transient properties of the extracellular environment during development.

Trophic or "Growth" Factors
A variety of macromolecular "factors" have been described that
affect neuronal development. Such factors affect the survival
and differentiation of their targets but, with one possible
exception (see next page), do not appear to stimulate cell

division. The first discovered and best characterized of these
is nerve growth factor or NGF (for reviews see (6, 18, 22, 30)).
NGF has a large number of different effects on its target cells,
sympathetic and sensory neurons, including maintenance of
survival, promotion of neurite outgrowth, tropic guidance of
neurite outgrowth, hypertrophy, and regulation of levels of
neurotransmitter-synthesizing enzymes. NGF's actions are thus
permissive (eg., survival) as well as modulatory (eg., regulation, but not specification of neurotransmitter synthesis).
The findings that NGF can cause otherwise round pheochromocytoma
(19) and adrenal medullary cells (2) to undergo neurite outgrowth suggests that NGF may also alter the phenotype of, and
thus have instructive actions on, target cells. Furthermore,
experiments in our laboratory indicate that NGF can either halt
mitosis (19) or stimulate proliferation (D.E. Burstein and
L.A. Greene, in preparation) of suitable targets. Hence, NGF
may have multiple developmental influences including regulation
of proliferation, entrance into the post-mitotic state, quali-
tative and quantitative regulation of differentiated properties,
prevention of death, and axonal guidance.

Tissue culture experiments have indicated the existence of
several macromolecular factors that appear to be distinct from
NGF: a) Ciliary ganglion and other non-NGF factors. Several
laboratories have described the ability of media conditioned
by several different cell types (eg., glia, heart, skeletal
muscle, iris) to promote the survival of and neurite outgrowth
from cultured ciliary ganglionic neurons (1, 9, 11, 12, 14, 24,
26, 32, 33, 52). At least several of these conditioned medium
preparations also maintain and promote neurite outgrowth by
cultured DRG and sympathetic neurons via factors that are
immunologically and biochemically distinguishable from NGF
(4, 5, 14, 23, 31, 32). Recent evidence suggests that separate
components in the conditioned media may independently regulate
the effects on survival and neurite outgrowth (1). b) Glial
factor. Medium conditioned by C6 glioma cells (37) or by cultured neonatal brain cells (46) has been reported to promote

morphological differentiation of cultured murine neuroblastoma
cells. c) CNS-neuron-affecting factors. Various tissue and
embryonic brain cell extracts have been reported to promote
survival of neurons in dissociated cultures of brain (49),
while medium conditioned by primary cultures of fetal or neonatal rat skeletal muscle, fibroblasts, or lung cells enhances
neuritic outgrowth from rat spinal cord explants (13).
d) Neurotransmitter synthesis. A factor present in the medium
of cultured glial cells and several other cell types "instructs"
cultured neonatal sympathetic neurons to synthesize acetylcholine and to suppress synthesis of catecholamines (for review
see (39)). Also, medium conditioned by cultured skeletal muscle
has been shown to greatly enhance the capacity of cultured
spinal cord cells to synthesize acetylcholine (16). It is noteworthy that these effects on neurotransmitter synthesis appear
to be distinct from effects on survival.

Substrate. The extracellular substrate has been considered to
be an important influence on neuronal development. In vitro experiments have indicated at least a clear permissive influence of
substratum on the quality and quantity of neurite outgrowtn (29).
While the extracellular substrate is likely to affect such processes as migration, definitive identification of the molecules
involved in such regulation has yet to be made.

WHAT ARE THE CHEMICAL PROPERTIES OF THE FACTORS?
The small molecular weight factors listed above (such as hormones) are structurally defined. Of the macromolecular factors,
NGF is the best characterized and has been studied in detail
(see (6, 18) for review). Briefly, mouse submaxillary NGF is
a protein of mol wt 27,000, consisting of two identical non-
covalently joined chains whose sequence has been determined.
This active protein can be isolated as part of a larger complex
and is derived by enzymatic cleavage from a "pro-NGF" species.
The non-NGF factors affecting survival and neurite outgrowth
by cultured neurons and cholinergic synthesis by cultured
sympathetic neurons have been partially characterized (12, 13,
23, 33, 39). These factors are macromolecules and appear to be

protein-like. The material affecting ciliary and other peripheral ganglia may have at least two separable components (1).
One of these appears to affect survival but not to stimulate
neurite outgrowth, while the other appears to promote neurite
outgrowth. The latter factor appears to attach to the polyornithine substrate on which the test cultures are maintained
and hence to be active in a non-soluble form (1, 9, 12). Since
this factor does not appear to promote outgrowth by target
neurons cultured on collagen (1), the possibility is raised
that part of its effect during development could be specified
by the local substratum. The above information notwithstanding,
it is still neither clear as to how many different macromolecular factors are present in the preparations employed by various groups nor the degree to which the different factors resemble one another.

BY WHAT ROUTES DO FACTORS REACH TARGETS?
Several possible routes of "delivery" of factors are available. a) Circulation. This is the most likely route of hormones and could also be at least one source for soluble macromolecular factors. Factors could also be supplied in limited
or gradient form via release into the local environment. b)
Transsynaptic. There is abundant evidence that NGF can be
taken up at nerve endings and retrogradely transported to dorsal root and sympathetic ganglia (for review see (6, 18, 22)).
It has been suggested that NGF could be transsynaptically transferred to neurons by target cells and then transported to the
cell body (22). c) Transcellular. As one possible example
of this mode, glia appear capable of synthesizing at least
several developmental factors, and it is conceivable that these
could be directly supplied to neurons. Precedence for glial-
neuronal transference exists in invertebrates (see Orkand, this
volume). d) Membrane bound. Developmentally influential signals could be bound to or exposed on the external membranes of
neighboring cells. A well documented example of this pertains
to stimulation of glial proliferation by certain axonal membranes (see below). e) Substrate bound. The substrate itself

or factors that bind (and possibly modify) the extracellular
substrate may influence development. An example of the latter
is the factor described above in conditioned media that, when
bound to polyornithine, promotes neurite outgrowth by cultured
ganglia.

WHAT IS THE TARGET SPECIFICITY OF THE CHEMICAL FACTORS?

One might expect a range of specificities with respect to targets for chemical factors influencing neuronal development.
Some factors could affect a wide variety of tissues as well as
a wide variety of different types of brain cells. An example
of this appears to be thyroxine. Other factors such as steroid
hormones may have a more restricted variety of targets both
within and outside of the nervous system. Thus, sex hormones
bind only to a specific subset of CNS cells including areas of
the thalamus, hypothalamus, and amygdala (for review, see (40)).
Additional factors may affect only a particular class of neurons.
The apparent presence of a factor or of factors affecting DRG,
sympathetic, and ciliary neurons (31) suggests the possibility
of a component active on neurons derived from the neural crest.
NGF shows great apparent specificity; its effects appear to be
normally restricted to sympathetic and certain sensory neurons.
Finally, our current notions of specificity of trophic interactions in the nervous system could also accommodate the possibility of factors that affect only a single type of neuron.
Some of the factors affecting parasympathetic ciliary neurons
may show such a restriction, but this possibility has not yet
been conclusively established.

IS RESPONSIVENESS TO CHEMICAL FACTORS TEMPORALLY LIMITED?

Neurons appear to be able to respond to at least some chemical
factors only during limited "windows" of development. For
instance, sex hormones and thyroxine have well-defined "critical periods" during which their presence or absence can alter
neuronal development (3, 20). The effects of NGF on DRG neurons
are present only during a discrete prenatal period (30). In
addition, at the developmental time that DRG neurons lose their

responsiveness to NGF, they appear to take on the requirement
for a different factor (4). As a final example, sympathetic
neurons appear to become increasingly refractive to the
"cholinergic factor" as they develop (see (39) for review).

In contrast, certain responses may be retained well beyond
development. For example, sympathetic neurons appear to maintain the capacity to respond to NGF even in adulthood. This
capacity could be correlated with the ability of sympathetic
neurons to regenerate their neurites.

WHAT REGULATES CRITICAL PERIODS OF RESPONSIVENESS TO CHEMICAL FACTORS?

Since most factors appear to work via specific receptors,
one possibility is that responsiveness is regulated by the
presence or absence of such receptors. There is evidence, for
example, that DRG neurons lose their specific receptors for
NGF at the time that they lose their responsiveness to and
requirement for the factor (see (30) for review). Such a
situation, on the other hand, may not hold for sex hormones
for which receptors are present in the adult CNS. In any case,
the question still remains as to what in turn regulates the
presence or absence of receptors.

IS THE PRESENCE OF CHEMICAL FACTORS TEMPORALLY LIMITED?

Clearly, there must be an "onset" before which developmental
factors are absent. The timing of this onset could in turn
temporally regulate neuronal development. For instance, in
the chick, the stages at which death of ciliary ganglion
neurons ceases is coincident with a large increase in ocular
neuronotrophic activity (26). What is less clear, in most
cases, however, is whether the presence or availabilities of
such factors are also limited to specific windows and, if so,
how the durations of such windows are in turn regulated. A
recent set of experiments (14) raises at least one possibility -
that of synaptic regulation of factor synthesis. Freshly
excised adult irides reportedly contain no detectable NGF

activity; however, NGF can be detected (and apparently synthesized) if this tissue is denervated or excised and cultured.
This capacity to synthesize NGF stops when the irides become
re-innervated.

HOW MANY DIFFERENT CHEMICAL FACTORS INFLUENCE THE DEVELOPMENT OF A GIVEN NEURON?

The answer to this question is of course not known, but for
peripheral neurons, responses appear to be present for multiple factors. For example, sympathetic neurons can be affected by NGF, conditioned medium factors, corticosteroids,
and the "cholinergic" factor. Moreover, these responses appear to co-exist at least during certain periods of development.

CAN ONE CHEMICAL FACTOR INFLUENCE RESPONSIVENESS TO ANOTHER FACTOR?

Several instances of such cross-regulation exist. There is
evidence that androgenization during development can alter
the responsiveness of the adult brain to estrogens (40). As
mentioned above, corticosteroids can either enhance the ability
of NGF to increase tyrosine hydroxylase synthesis or antagonize
the capacity of NGF to promote neurite outgrowth. Also, NGF
has been reported to down-regulate epidermal growth factor (EGF)
receptors in cultured PC12 pheochromocytoma cells (25). This
finding raises the provocative possibility that a "differentiation" factor could affect cell proliferation by regulating responses to other mitogenic factors.

WHAT IS THE TISSUE SOURCE OF THE CHEMICAL FACTORS?

Are specific factors supplied by specific tissues? Do target
organs specifically secrete large amounts of specific factors
that have trophic and tropic effects? How is factor synthesis
and release regulated? Do glial cells supply specific factors
to nearby neurons?

Hormones such as thyroxine, corticosteroids, and testosterone
are supplied by discrete, identifiable tissues and are

distributed systemically. With respect to macromolecular
factors, the issue is less clear. The most studied of these,
NGF, while synthesized in large amounts by certain adult
tissues (e.g., adult male mouse submaxillary gland and guinea
pig prostate), does not presently have an identified developmental source. The factor has been found to be synthesized
in small quantities by a variety of cultured cells including
glial cell lines, fibroblasts, skeletal muscle, and adrenal
cells (see (6, 18, 22) for review). There is no good evidence,
however, that tissues innervated by NGF-responsive neurons
show a differential capacity to synthesize, store, or release
the factor. One possible scenario consistent with present
information is that during developmental periods before extensive neurite outgrowth, NGF is supplied by a variety of
tissues via the circulation. During the limited temporal
window in which specific synaptic connections are formed, the
factor could be supplied locally in large amounts by target
organs and thus serve a tropic as well as trophic function.
Once innervation takes place, as in the iris experiments cited
above, synthesis would again decrease. Thus, it may be that
careful time studies will be necessary to pinpoint the developmental sources of NGF. With regard to the non-NGF macromolecular factors mentioned above, there is some evidence for
preferential synthesis by peripheral targets or glial cells.
However, in each case, the presently available data do not
seem to be extensive or consistent enough to draw any definitive conclusions on this point. It is anticipated that the
search for selective sources (if any) for such factors will
be materially aided once they are isolated and better characterized.

Little is known about regulation of synthesis of the developmentally influential factors. NGF synthesis is clearly regulated by testosterone in certain tissues (see (17) for review),
and there is evidence that the levels of this factor in certain
glial cell lines can be affected by glucocorticoids (47).

Another possible source of developmental factors to be considered in mammals is the mother. Several maternal hormones
pass the placental barrier and could affect neuronal differentiation. Also, there is evidence that NGF is present in
the human placenta (see (22) for review) and therefore might
be available to the fetus.

BY WHAT MECHANISMS OF ACTION DO THE FACTORS WORK?
The primary mechanisms of action of hormones such as thyroxine and steroids are apparently the same in the nervous system as they are in other tissues. For example, steroid hormones in brain interact with cytosolic and nuclear receptors.
This in turn leads, by mechanisms that are not yet well understood, to regulation of specific genes (see (35, 40) for
review). Even less is known about the mechanisms of actions
of macromolecular factors. Do they interact with specific
receptors? What is the cellular locus of such receptors?
Are several receptors involved? Are secondary messengers or
transductive mechanisms involved? Do the factors work at
the level of the genome? Thus far, only for NGF are some
insights available (for review see (6, 18)). In this case,
specific high affinity receptors are present on the surfaces
and also apparently in or on the nucleus of target cells.
The importance of the nuclear binding site is unclear, but
the observation that NGF is internalized by its targets suggests that such receptors may have a functional role. Secondary messengers for NGF, if any, have not yet been identified. Experiments with cultured pheochromocytoma cells have
revealed that NGF has separate transcription-dependent and
transcription-independent actions and that each of these types
of actions is required for initiation of neurite outgrowth.
A "priming" model has been proposed in which the transcription
dependent-action causes synthesis and accumulation of material
(presently unidentified) that is required for neurite outgrowth. In this model such material would lead to neurite outgrowth, but only if NGF were present and working in a non-
transcriptional mode (17). The different pathways of NGF's
action could in part account for its ability to promote a
number of different types of responses by its target cells.

CHEMICAL INFLUENCES ON GLIAL DEVELOPMENT

The questions posed above for neurons can for the most part
also be posed for glia. For reasons of space and scope and
since this subject has recently been well reviewed ((7), see
Bunge, this volume), only a short discussion can be given here.
Several influences on glial development have thus far been
uncovered. Studies with avian (21, 34) and rodent (7, 43-45)
culture systems have revealed that sympathetic and sensory
neurons can stimulate proliferation of Schwann cells. This
mitogenic activitiy is present on the neurites and appears to
be membrane-bound, surface-exposed, and trypsin-sensitive
(43-45). The mitogenic activity thus appears to require direct neuronal-glial contact. The implications of such findings
are that during development, the axonal signal could stimulate
Schwann cell proliferation; once the neuron is invested with a
glial sheath, the signal would no longer be accessible. Also,
if injury were to cause unsheathment of the axon, the mitogenic
signal would once again be exposed (7). There is also evidence
that a similar type of mitogenic signal may be present on CNS
neurons (21). Factors causing Schwann cell proliferation have
also been found in extracts of brain and pituitary (41). The
active components appear to be proteins that are distinct from
previously characterized growth factors. Another possible
influence on glial proliferation is the well-known epidermal
growth factor (EGF). EGF has been reported to stimulate proliferation of cultured human glial cells (54). Recent experiments suggest that the glial cells affected have the morphology
of oligodendrocytes and that such effects of EGF may occur
only during a very limited temporal window (E. Trenkner, personal communication). The mechansims of action of the above
factors are not clear, but there is limited evidence that glial
proliferation can be affected by at least two routes, one of
which involves an increase in intracellular cyclic AMP (41).
An additional influence on glial development that has been
raised is extracellular substrate. It has been reported that
collagen, but not certain other substrates, is permissive for
formation of myelin in vitro (7).

REFERENCES

(1) Adler, R., and Varon, S. 1980. Cholinergic neurotrophic
     factors: V. Segregation of survival- and neurite-promoting
     activities in heart-conditioned medium. Brain Res. 188:
     437-448.

(2) Aloe, L., and Levi-Montalcini, R. 1979. Nerve growth
     factor-induced transformation of immature chromaffin cells
     in vivo into sympathetic neurons: Effect of antiserum to
     nerve growth factor. Proc. Natl. Acad. Sci. USA 76:
     1246-1250.

(3) Balazs, R. 1974. Influences of metabolic factors on
     brain development. Br. Med. Bull. 30: 126-134.

(4) Barde, Y.A.; Edgar, D.; and Thoenen, H. 1980. Sensory
     neurons in culture: changing requirements for survival
     factors during embryonic development. Proc. Natl. Acad.
     Sci. USA 77: 1199-1203.

(5) Barde, Y.A.; Lindsay, R.M.; Monard, D.; and Thoenen, H.
     1978. New factor released by cultured glioma cells sup-
     porting survival and growth of sensory neurones. Nature
     274: 818.

(6) Bradshaw, R.A. 1978. Nerve growth factor. Ann. Rev.
     Biochem. 47: 191-216.

(7) Bunge, R.P.; Bunge, M.B.; and Cochran, M. 1978. Some
     factors influencing the proliferation and differentiation
     of myelin forming cells. Neurology 28: 59-67.

(8) Chalazonitis, A., and Fischbach, G.D. 1980. Elevated
     potassium induces morphological differentiation of dorsal
     root ganglionic neurons in dissociated cell culture. Dev.
     Biol. 78: 141-151.

(9) Collins, F. 1978. Induction of neurite outgrowth by a
     conditioned-medium factor bound to the culture substratum.
     Proc. Natl. Acad. Sci. 75: 5210-5213.

(10) Cone, C.D., Jr., and Cone, C.M. 1976. Induction of
     mitosis of mature neurons in central nervous system by
     sustained depolarization. Science 192: 155-158.

(11) Coughlin, M.D. 1975. Target organ stimulation of para-
     sympathetic nerve growth in the developing mouse subman-
     dibular gland. Dev. Biol. 43: 140-158.

(12) Coughlin, MD.; Bloom, E.E.; and Black, I.B. Character-
     ization of a neuronal growth factor from mouse heart cell
     conditioned medium. Dev. Biol., in press.

(13) Dribin, L.B., and Barrett, J.N. 1980. Conditioned
      medium enhances neurite outgrowth from rat spinal cord
      explants. Dev. Biol. 74: 184-195.

(14) Ebendal, R.; Olson, L.; Seiger, A.; and Hedlund, K.-O.
      1980. Nerve growth factors in the rat iris. Nature
      286: 25-28.

(15) Eränkö, O.; Eränkö, L.; Hill, C.E.; and Burnstock, G.
      1972. Hydrocortisone-induced increase in the number of
      small intensely fluorescent cells and their histochemically
      demonstrable catecholamine content in cultures of sympathe-
      tic ganglia of the newborn rat. Histochem. J. 4: 49-58.

(16) Giller, E.L., Jr.; Neale, J.H.; Bullock, P.N.; Schrier,
      B.K.; and Nelson, P.G. 1977. Choline acetyltransferase
      activity of spinal cord cell cultures increased by co-
      culture with muscle and by muscle-conditioned medium.
      J. Cell Biol. 74: 16-29.

(17) Greene, L.A., Burstein, D.E., and Black, M.M. 1980.
      The priming model for the mechanism of action of nerve
      growth factor: Evidence derived from clonal PC12 pheo-
      chromocytoma cells. In Tissue Culture in Neurobiology,
      eds. E. Giacobini, A. Vernadakis and A. Shahar, pp.
      313-319. New York: Raven Press.

(18) Greene, L.A., and Shooter, E.M. 1980. The nerve growth
      factor: Biochemistry, synthesis, and mechanism of action.
      Ann. Rev. Neurosci. 3: 353-402.

(19) Greene, L.A., and Tischler, A.S. 1976. Establishment of
      a noradrenergic clonal line of rat adrenal pheochromocytoma
      cells which respond to nerve growth factor. Proc. Natl.
      Acad. Sci. USA 73: 2424-2428.

(20) Hamburgh, M.; Mendoza, L.A.; Burkart, J.F.; and Weil, F.
      1971. The thyroid as a time clock in the developing
      nervous system. In Cellular Aspects of Neural Growth and
      Differentiation, pp. 321-328. Los Angeles: University
      of California Press.

(21) Hanson, G., and Partlow, L. 1977. Neuronal stimulation
      of (^{3}H) thymidine incorporation by non-neuronal cells.
      Soc. Neurosci. Abstr. 3: 535.

(22) Harper, G.P., and Thoenen, H. 1980. Nerve growth factor:
      Biological significance, measurement, and distribution.
      J. Neurochem. 34: 5-16.

(23) Helfand, S.L.; Riopelle, R.J.; and Wessels, N.K. 1978.
      Non-equivalence of conditioned medium and nerve growth
      factor for sympathetic, parasympathetic and sensory
      neurons. Exp. Cell Res. 113: 39-45.

318   L.A. Greene

(24) Helfand, S.L.; Smith, G.A.; and Wessels, N.K. 1976.
      Survival and development in culture of dissociated para-
      sympathetic neurons from ciliary ganglia. Dev. Biol.
      50: 541-547.

(25) Huff, K.R., and Guroff, G. 1979. Nerve growth factor-
      induced reduction in epidermal growth factor responsive-
      ness and epidermal growth factor receptors in PC12 cells:
      An aspect of cell differentiation. Biochem. Biophys.
      Res. Comm. 89: 175-180.

(26) Landa, K.B.; Adler, R.; Manthrope, M.; and Varon, S.
      1979. Cholinergic neuronotrophic factors. III. Develop-
      ment increase in trophic activity for chick embryo ciliary
      ganglion neurons in their intraocular target tissues.
      Dev. Biol. 74: 401-408.

(27) Lanier, L.P.; Dunn, A.J.; and Van Hartesveldt. 1976.
      Development of neurotransmitters and their function in
      brain. In Reviews of Neuroscience, eds. S. Ehrenpreis
      and I.J. Kopin, vol. 2, pp. 195-256. New York: Raven
      Press.

(28) Lasher, R.S., and Zagon, I.S. 1972. The effect of
      potassium on neuronal differentiation in cultures of
      dissociated newborn rat cerebellum. Brain Res. 41:
      482-488.

(29) Letourneau, P.C. 1977. Regulation of neuronal morpho-
      genesis by cell-substratum adhesion. Soc. Neurosci.
      Symp. 1: 67-81.

(30) Levi-Montalcini, R., and Angeletti, P.U. 1968. Nerve
      growth factor. Physiol. Rev. 48: 534-569.

(31) Lindsay, R.M. 1979. Adult rat brain astrocytes support
      survival of both NGF-dependent and NGF-insensitive
      neurones. Nature 282: 80-82.

(32) Lindsay, R.M., and Tarbit, J. 1979. Developmentally
      regulated induction of neurite outgrowth from immature
      chick sensory neurons (DRG) by homogenates of avian or
      mammalian heart, liver and brain. Neurosci. Lett.
      12: 195-200.

(33) Manthorpe, M.; Skaper, S.; Adler, R.; Landa, K.; and
      Varon, S. 1980. IV. Fractionation properties of an
      extract from selected chick embryo tissues. J. Neurochem.
      34: 69-78.

(34) McCarthy, K., and Partlow, R. 1976. Neuronal stimulation
      of tritiated thymidine incorporation by primary cultures
      of highly purified non-neuronal cells. Brain Res. 114:
      415-426.

(35) McEwen, B.S.; Davis, P.G.; Parsons, B.; and Pfaff, D.W. 1979. The brain as a target for steroid hormone action. Ann. Rev. Neurosci. 2: 65-112.

(36) McLennan, I.S.; Hill, C.E.; and Hendry, I.A. 1980. Glucocorticosteroids modulate transmitter choice in developing superior cervical ganglion. Nature 283: 206-207.

(37) Monard, D.; Solomon, F.; Kentsch, M.; and Gysin, R. 1973. Glia-induced morphological differentiation in neuroblastoma cells. Proc. Natl. Acad. Sci. USA 70: 1682-1687.

(38) Otten, U., and Thoenen, H. 1976. Modulatory role of glucocorticoids on NGF-mediated enzyme induction in organ cultures of sympathetic ganglia. Brain Res. 111: 438-441.

(39) Patterson, P.H. 1978. Environmental determination of autonomic neurotransmitter function. Ann. Rev. Neurosci. 1: 1-17.

(40) Plapinger, L., and McEwen, B.S. 1978. Gonadal steroid-brain interactions in sexual differentiation. In Biological Determination of Sexual Behavior, pp. 153-218. New York: Wiley.

(41) Raff, M.C.; ABney, E.; Borckes, J.; and Hornby-Smith, A. 1978. Schwann cell growth factors. Cell 15: 813-822.

(42) Raisman, G., and Field, P.M. 1973. Sexual dimorphism in the neuropil of the preoptic area of the rat and its dependence on neonatal androgen. Brain Res. 54: 1-29.

(43) Salsezer, J.L., and Bunge, R.P. 1980. Studies of Schwann cell proliferation. I. Analysis in tissue culture of proliferation during development, Wallerian degeneration and direct injury. J. Cell Biol. 84: 739-752.

(44) Salsezer, J.L.; Bunge, R.P.; and Glaser, L. 1980. Studies of Schwann cell proliferation. III. Evidence for the surface localization of the neurite mitogen. J. Cell Biol. 84: 767-778.

(45) Salsezer, J.L.; Williams, A.K.; Glaser, L.; and Bunge, R.P. 1980. Studies of Schwann cell proliferation II. Characterization of the stimulation and specificity of the response to a neurite membrane fraction. J. Cell Biol. 84: 753-766.

(46) Schurch-Rathgeb, Y., and Monard, D. 1978. Brain development influences the appearance of glia factor-like activity in rat brain primary cultures. Nature 273: 308-309.

(47)   Schwartz, J.P., and Costa, E. 1978. Regulation of
       nerve growth factor content in a neuroblastoma cell
       line. Neurosci. 3: 473-480.

(48)   Scott, B.S., and Fisher, K.C. 1970. Potassium concen-
       tration and number of neurons in cultures of dissociated
       ganglia. Exp. Neurol. 27: 16-22.

(49)   Sensenbrenner, M. 1977. Dissociated brain cells in
       primary cultures. In Cell, Tissue and Organ Cultures
       in Neurobiology, eds. S. Federoff and L. Hertz, pp.
       191-213. New York: Academic Press.

(50)   Unsicker, K.; Krisch, B.; Otten, U.; and Thoenen, H.
       1978. Nerve growth factor-induced fiber outgrowth from
       isolated rat adrenal chromaffin cells: impairment by
       glucocorticoids. Proc. Natl. Acad. Sci. USA 75:
       3498-3502.

(51)   Varon, S. 1975. Nerve growth factor and its mode of
       action. Exp. Neurol. 48: (part 2) 75-92.

(52)   Varon, S.; Manthorpe, M.; and Adler, R. 1979. Cholin-
       ergic neurotrophic factors: I. Survival, neurite out-
       growth and choline acetyltransferase activity in mono-
       layer cultures from chick embryo ciliary ganglia. Brain
       Res. 173: 29-45.

(53)   Wood, P., and Bunge, R. 1975. Evidence that sensory
       axons are mitogenic for Schwann cells. Nature 256:
       662-664.

(54)   Westermark, B. 1976. Density dependent proliferation
       of human glia cells stimulated by epidermal growth factor.
       Biochem. Biophys. Res. Comm. 69: 304-310.

(55)   Wurtman, R.J., and Axelrod, J. 1966. Control of enzy-
       matic synthesis of adrenaline in the adrenal medulla by
       adrenal cortical steroids. J. Biol. Chem. 241:
       2301-2305.

Neuronal-glial Cell Interrelationships, ed. T.A. Sears, pp. 321-336.
Dahlem Konferenzen 1982. Berlin, Heidelberg, New York: Springer-Verlag.

Developmental Expression of Antigenic Markers in Glial Subclasses

M. Schachner, I. Sommer, C. Lagenaur, J. Schnitzer
Dept. of Neurobiology, University of Heidelberg
6900 Heidelberg, F. R. Germany

Abstract. Several antibodies will be described which are directed to the cell surfaces of oligodendrocytes and intracellular antigens of astrocytes and astroglial subclasses.

The four antigens (O1 to O4), recognized by monoclonal antibodies
on the surface of oligodendrocytes can be subdivided into two
groups by their sequence of developmental expression. Most, if
not all oligodendrocytes of the more mature subgroup as identified by their expression of O1 and O2 antigens also express
galactocerebroside. The second group, identified by the presence of O3 and O4 antigens and the absence of galactocerebroside and O1 and O2 antigens generally shows a less elaborate
morphology. The percentage of cells which express at the same
time both O3 and O4 antigens as well as galactocerebroside increases with time in culture, thus providing indirect evidence
that maturation is achieved in vitro.

Intracellular antigens of astroglial subclasses are recognized
by two monoclonal antibodies: C1 and M1. These subclasses do
not correspond to the classical ones of protoplasmic and fibrillary astrocytes. In adult mice C1 antibody detects ependymal cells and radial glial fibers in cerebellar cortex and retina. M1 antibody reacts with astrocytes localized in cerebellar
white matter and other brain regions including white and gray
matter rich areas. While C1 antigen is already detectable in
radial fibers at embryonic day 11, the earliest stage tested
so far, M1 antigen is not expressed before the end of the first
postnatal week. Both antigens are expressed not only in the
astrocyte subclass that they are confined to at later ages, but
also transiently in other astrocyte types. Several neurological
mutants of the mouse, i.e., Purkinje cell degeneration (pcd),
reeler (rl), staggerer (sg), and weaver (wv), show abnormal expression of C1 and M1 antigens in the cerebellar cortex but

M. Schachner et al.

appear normal in the retina. While C1 antigen remains detectable in ependymal cells in these mutants, it is not apparent
in Bergmann glia. In contradistinction, M1 antigen, which is
restricted to astrocytes in cerebellar white matter in normal
littermates, additionally appears in Bergmann glia in these
mutants.

Vimentin, the subunit protein of intermediate-sized filaments,
is present in all astrocytes and ependymal cells in the adult
neuroectoderm and in radial fibers and ventricular cells in
the embryonic brain. Vimentin expression appears unaffected
by neurological mutations in cerebellum and retina.

INTRODUCTION

The ontogeny of the different types of neural cells and the
developmental relationships between these cells pose a major
problem in neurobiology and developmental biology in general.
All classes of neurons and glia presumably arise from a common stem cell, the germinal cell (12), which constitutes the
epithelium of the neural tube. It is not known when the program of differential gene expression is initiated in these
germinal cells, which is presumed to result in the divergent
differentiation pathways and cell lineages of various types
of neurons and glia.

The origin of glia seems particularly mysterious at present,
since in contrast to neurons, they continue to proliferate and
divide often throughout life (24,40). Whereas different types
of neurons originate in distinct temporal patterns by with-
drawing from the cell cycle of their neuroblast precursors, an
orderly sequence of growth and differentiation of glia and
glial subpopulations has yet to become apparent.

Equally unsolved is the problem of differentiative pathways
taken by the glial subclasses, particularly the astroglia and
oligodendroglia. These two types of macroglia seem to arise
from a common precursor, the glioblast (32,33). Whether these
are multipotential glial cells which are capable of giving
rise to astroglia as well as oligodendroglia (44,45), or whether they constitute direct subsets of glial precursors is presently undecided. It is also an open question, whether

differentiated astrocytes and oligodendrocytes arise simultaneously or consecutively (40,42), but it seems widely accepted
that oligodendrocytes differentiate later than astrocytes and
do not arise by transformation of intermediate forms (29).

The problem of origin and differentiation of glial cells has
been hampered both by the lack of understanding of glial functions and by the fact that unequivocal criteria for identification of glia at early developmental stages have not been
established. In addition to the existing ultrastructural description of developing glia, a search for biochemical markers
detectable by appropriate immunological reagents seems imperative. These marker molecules may also be instrumental in analyzing cell lineage relationships among immature precursor
cells for which morphological criteria alone are insufficient.
Also, with cells in vitro where characteristic morphology can
be grossly altered, these markers are useful for unequivocal
cell identification. Since cellular differentiation is generally accompanied by the appearance of molecules that are
related to the performance of the cell's special tasks, a
characterization of such molecules might aid in the recognition of the cell's unique functional properties.

As a major subclass of neural cells, astroglia have been known
for more than a century as a neural population distinct from
neurons and other glial cell types by various structural and
functional features. Their functional role in cell-cell interactions in the nervous system has, however, remained largely
obscure. They may provide nutritive and structural support of
neurons and may be implicated in both the chemical and the
physical isolation of neurons and their synapses (24). They
have been shown to retain the ability to divide and grow in
response to neural damage (13,19,43). Radial astroglia in the
developing brain have been proposed to act as guides for migrating neurons in the cerebellar and cerebral cortices (26,34).

Indeed, a defect in Bergmann glial cells (Golgi epithelial cells) of the cerebellar cortex has been suggested as the cause of the granule cells' inability to migrate successfully during cerebellar development of the weaver mutant mouse (27).

Historically, astroglia have been divided into two subclasses. Astrocytes in white matter regions abound in glial filaments and have therefore been designated fibrillary or fibrous astrocytes (24). Protoplasmic astrocytes with more attenuated processes are typical for gray matter brain regions and express characteristically fewer glial fliaments (24). At present it is not known whether fibrillary or protoplasmic astrocytes display distinct functional traits in vivo or in vitro. Both fibrillary and protoplasmic astrocytes are characterized by the expression of glial fibrillary acidic protein which is found in astroglia but in no other neural and nonneural cell types (1). This protein of apparent molecular weight 54,000 is the presumed monomeric subunit of the 100-$\overset{\circ}{A}$ glial filament system (4,5,8). Its function is presently unknown.

At a particular stage during development, the differentiative pathway of oligodendrocytes diverges from the one taken by astrocytes. Oligodendrocytes then appear to undergo morphologically distinct transformation from immature to mature forms (14,21,30) with their capability to myelinate axons (2,28). Myelination is the last in these differentiative steps and possibly the most complex. The cellular mechanisms of proliferation and differentiation of oligodendrocytes, particularly their intercellular communication with nerve axons, have remained largely obscure.

The counterparts of oligodendrocytes in the peripheral nervous system have been termed Schwann cells. In contrast to oligodendrocytes in the central nervous system, which are of neuroectodermal origin, Schwann cells arise from the neural crest (6,10,11), migrate over longer distances to the peripheral nerves (46), proliferate, and finally differentiate either into myelin forming cells or ensheathing cells which do not generate

myelin. Despite their differences in origin, oligodendrocytes
and Schwann cells are endowed with some basically similar structural and functional properties.

Imperative for the recognition of subtle molecular differences
and similarities among neural cell classes is the availability
of refined immunological methodology. Most neural antigens have
so far been defined predominantly by xenogeneic antisera for
which species differences exist between the animal yielding tissue or cells for immunization and the immunized host animal. To
render the resulting antisera specific for nervous tissue and/or
neural cell type, antibodies to species-specific antigens which
are not of interest have to be removed by extensive absorption
with non-neural tissues. Even though operationally specific
antisera could thus be obtained, they are most likely still
heterogeneous with respect to antigenic specificities, antibody
subclasses, and binding affinities.

Monoclonal antibodies which can be obtained with relative ease
by the myeloma-hybrid technique of Köhler and Milstein (15,16)
circumvent these various problems of conventional antisera.
Especially in cases when complex mixtures of antigens are used
for immunization monoclonal antibodies are able to dissect this
heterogeneity and potentially enable the detection of minor
antigens and minor cell populations when appropriate assay systems are used. When successful, this approach would open new
avenues in the investigation of neural cell surfaces and the
characterization of more refined cell surface differences among
developing and mature cell types and topographically outstanding
specializations at the cell surface.

RESULTS
C1 Antigen (Sommer et al., submitted for publication)
A monoclonal antibody designated anti-C1 was obtained from a
hybridoma clone isolated from a fusion of NS1 myeloma with
spleen cells from BALB/c mice injected with homogenate of white
matter from bovine corpus callosum.

In the adult mouse nervous system, C1 antigen is detectable by
indirect immunohistology in the processes of Golgi epithelial
cells in the cerebellum and of Müller cells in the retina,
whereas other astrocytes that express glial fibrillary acidic
protein in these brain areas are negative for C1. In addition,
C1 antigen is expressed in most, if not all ependymal cells
and in larger blood vessels, but not in capillaries. In the
developing, early postnatal cerebellum, C1 antigen is not confined to Golgi epithelial cells and ependymal cells, but is
additionally present in astrocytes of presumptive white matter
and Purkinje cell layer. In the embryonic neuroectoderm, C1
antigen is already expressed at day 11, the earliest stage
tested so far. The antigen is distinguished in radially oriented structures in telencephalon, pons, pituitary, and retinal
anlage. Ventricular cells are not labeled by C1 antibody at
this stage.

C1 antigen is not detectable in astrocytes of adult cerebella
from the neurological mutant mice staggerer, reeler, and weaver
but is present in ependymal cells and larger blood vessels.

C1 antigen is expressed not only in the intact animal but also
in cultured cerebellar astrocytes and fibroblast-like cells.
It is localized intracellularly.

M1 Antigen (18)
A monoclonal antibody was detected that distinguishes astrocyte subclasses in mouse cerebellum. This antibody, designated
anti-M1, is a product of a hybridoma that arose from the fusion of NS1 myeloma cells and splenocytes derived from a rat
immunized with crude membranes from early postnatal mouse cerebella.

In fresh frozen sections of adult mouse cerebellum, M1 antibody
stained by indirect immunofluorescence only appeared in astrocytes in the white matter, but not in granular or molecular
layers. Double-immunolabeling techniques using GFA protein

antiserum to mark astrocytes revealed the identity of astrocytes in white matter and proved, by successful staining of
astrocytes in granular and molecular layers, that antigenic
structures were indeed accessible.

At earlier stages of development, M1 was first detectable at
about postnatal day 7 in white matter astrocytes. At postnatal
day 10, M1 antigen was additionally detectable in Bergmann glial
fibers, a special type of cerebellar astrocyte, and in astrocytes of the granular layer. M1 antigen expression in these
latter types of astrocytes was found to be transient and could
not be detected after the fourth postnatal week, when the general maturation of the mouse cerebellum has reached completion.

In monolayer cultures of early postnatal cerebella, M1 antigen
was detected in a subpopulation of the glial fibrillary acidic
protein-positive astrocytes. M1 seems to be an intracellular
antigen, since it was detectable only in fixed cultured cells
which allow intracellular penetration of the antibody. Interestingly, cerebella of adult neurological mice, weaver and
staggerer (see (3) for review), showed abnormal persistence
of M1 antigen in Bergmann glial fibers, resulting in an immature
phenotype of antigen expression. The developmental regulation
of M1 expression and the abnormal expression in neurological
mutations affecting the cerebellum suggest that M1 may be a
useful indicator of astroglial differentiation and a marker for
subpopulations of astrocytes.

O Antigens (O1 to O4) (31,41)
Four monoclonal antibodies are characterized that have been
obtained from a fusion of mouse myeloma NS1, with spleen cells
from BALB/c mice immunized with white matter from bovine corpus callosum. The corresponding antigens (O antigens) are designated O1, O2, O3, and O4.

The localization of these antigens was investigated by indirect
immunofluorescence in cultures of early postnatal mouse cerebellum,

328   M. Schachner et al.

cerebrum, spinal cord, optic nerve, and retina. When tested
on live cultures, none of the O antibodies reacted with the
surface of astrocytes, neurons, or fibroblasts, however, all
are positive on the surface of oligodendrocytes. The identity
of these cells was determined by double immunolabeling experiments with independent cell-type-specific antigenic markers
(glial fibrillary acidic protein, tetanus toxin receptors, fibronectin, and galactocerebroside). Antigens O1 and O2 are expressed on galactocerebroside-positive cells, whereas O3 and O4
are expressed on additional cells that are negative for any of
the markers tested.

Evidence will be given that these O1 antigen-negative, but O3
or O4 antigen-positive cells, are direct precursors to O1 antigen, galactocerebroside, or NS1 antigen-positive cells by in-
vestigating the developmental sequence of antigen expression.

On freshly trypsinized viable single cerebellar cell suspensions,
O1 and O2 antigens were not detectable at birth, whereas O3 and
O4 were present. At postnatal day 7 all O antigens were readily
detectable. In monolayer cultures of fetal mouse pons and cerebellum, expression of antigens O3 and O4 preceded that of galactocerebroside, NS1, O1, and O2 antigens. In these cultures O3
and O4 antigens were expressed by more simple or immature looking
cells, while O1 and O2 antigen-positive cells were found on more
elaborate and amply process-bearing cells. After longer culture
periods, more O3 and O4 antigen expressing cells also expressed
galactocerebroside and O1 antigen.

Direct evidence that O1 antigen bearing cells are derived from
O1 negative, O4 positive cells has come from two experiments.
In one, O1 antigen-positive cells were killed by complement-
dependent immunocytolysis in single cell suspensions from
freshly trypsinized cerebellum. Residual O4 antigen-positive
cells were isolated by using magnetic beads which had been covalently coupled to O4 antibody. The isolated O4 positive,
but O1 negative, cells were maintained in culture and O1 antigen

expression monitored by indirect immunofluorescence. O1 antigen became detectable after one day in vitro (D. Meier, C.
Lagenaur, and M. Schachner, unpublished results). In the second
experiment, O1 negative, but O4 positive cells obtained as described above were labeled with rhodamine-conjugated antibodies
to O4 antigen at the time of plating in culture. The appearance
of O1 antigen was detected by fluorescein-conjugated O1 antibody. After one day in vitro, cells which had started to inter-
nalize the rhodamine label had become positive for O1 antibody
(I. Sommer and M. Schachner, unpublished results).

To investigate whether other regions of the central nervous
system also expressed the O antigens, retina, cerebrum, optic
nerve, and spinal cord cultures were reacted with O antibodies.
None of the O antigens were found expressed in live cultures of
neural mouse retina known to lack oligodendrocytes, but were
detectable on oligodendrocytes of other parts of the central
nervous system (cerebrum, spinal cord, and optic nerve). In
fixed monolayer cultures, where intracellular antigens are accessible to antibody, O1, O2, and O3 antibodies label astrocytes
in a GFA protein-like pattern. Labeling of astrocytes was, however, never seen on unfixed live cultures. The antibodies can
therefore be considered specific for the cell surface of oligodendroglia.

O antigens were found not only in mouse, but also in rat, chick-
en, and human central nervous systems. All four antibodies be-
long to the IgM immunoglobulin subclass and have been used in
complement-dependent cytotoxic elimination of oligodendrocytes
in culture. At limiting antibody dilutions processes of oligodendrocytes were lysed in preference to cell bodies.

In the peripheral nervous system, in 2 to 7-day-old cultures
of neonatal dorsal root ganglion cells, O1 and O2 antigens are
hardly detectable under the culture conditions used, whereas
antigens O3 and O4 are found on some but not all Schwann cells.
After 7 days in vitro fewer Schwann cells are positive for O3
and O4 antigens than at 2 days.

Vimentin (35)

Vimentin is the protein of a type of intermediate-sized (7-11nm)
filament found in certain tissues, especially mesenchymal ones
of vertebrates (9).

Vimentin distribution was studied in histological sections and
monolayer cultures of mouse nervous system. To identify cell
types and correlate staining pattern, double immunolabeling experiments were performed using the markers described above for
identification of the various neural cell populations.

In the adult mouse nervous system, vimentin is expressed in
astrocytes and ependymal cells but not in neurons or oligodendrocytes. Specific labeling of astrocytes, but not of any other
neural cell type, is also evident in primary cultures of early
postnatal mouse cerebellum. In these cultures fibronectin-
positive fibroblasts or fibroblast-like cells are also positive
for vimentin.

Expression of vimentin has already been found in 11-day-old
mouse embryos, the earliest stage tested so far. In histological sections, vimentin-positive radially oriented fibers span
the neural tube from the ventricular lumen to the outer surface. Cells lining the ventricular lumen, i.e., the ventricular cells, are also vimentin positive. At this early stage
vimentin does not show an easily visible overlap of immunolabeling patterns with neurofilament protein. In cultures of
cerebellar cells from 11-day-old telencephalon maintained for
three days in vitro, vimentin is expressed in fibronectin-
positive fibroblasts or fibroblast-like cells but not in neurons
expressing tetanus toxin receptors. At this early stage, GFA
protein is not positive in astroglial-like cells that already
express vimentin but becomes expressed in these cultures after
longer culture periods.

In the neurological mutants Purkinje cell degeneration, reeler,
staggerer, and weaver, vimentin is expressed in all GFA protein-
positive astrocytes and, additionally, in all ependymal cells.

After treatment of cerebellar cultures with colcemid a characteristic aggregation of vimentin-positive filaments in perinu-
clear rings and whorls can be seen.

CONCLUSIONS

The present studies have shown that monoclonal antibodies can
distinguish among subclasses of astroglia cells and define a
population of oligodendrocytes which is less differentiated than
the one characterized by presently available markers. These
refined destinctions have hitherto not been possible using con-
ventionally prepared polyclonal antibodies. Furthermore these
antibodies resulted from immunizations using a very heterogeneous
mixture of antigens in form of homogenates and crude particulate
fractions of whole brain parts showing that purified antigens
or antigen mixtures of limited heterogenity are not a prerequisite for a successful hybridoma approach to the nervous system.

In particular, the unique features of C1 antigen have allowed
inferences with regard to developmental and functional relationships between ependyma, astrocytic subpopulations (comprising
Bergmann glia and Müller cells with radially oriented processes,
but not radial astroglia in the hippocampus), and primitive
radial glial fibers. Radial glia in the cerebellum and retina
are direct descendants of the primitive radial glia, the processes of which span the neural tube from the ventricular lumen
to the surface. While Müller cells remain attached to the outer
and inner surfaces of the neuroectoderm at adult stages, Bergmann glia lose their connection to the inner surface to retain
solely their apposition to the pial surface. The radial fibers
of astrocytes in the area dentata of the hippocampus arise after
the disappearance of the primitive radial glial processes (A.
Privat, personal communication) as a result of a subsequent
elaboration of secondary processes, an event which has been
described in the formation of astrocytes in the occipital lobe
of the monkey (34). It seems plausible, therefore, to speculate
that direct descendants of primitive radial glia retain features
in common to their precursors and the C1 antigen is an indicator
of such a relationship.

It remains unsolved why Bergmann glia are repressed in C1 antigen expression in several neurological mouse mutants affecting
development of the cerebellum in different ways (see (3), for
review). A mutant in which the interaction of Bergmann glia
and the migrating granule cell neurons is disturbed (weaver) lacks
C1 expression at adult stages as do Purkinje cell degeneration
(a mutant which has lost all Purkinje cells by the fifth postnatal week), staggerer (which loses granule cell neurons and
possesses abnormal Purkinje cells), or reeler (in which the
laminar structure of cortical layers is disturbed). Interestingly, ependyma, the cells lining the ventricular lumen of adult
animals, are not affected by the mutations but express C1 antigen
normally.

Expression of C1 antigen in ependymal cells would indicate that
these share common properties with astrocytes, as suggested by
several authors using ultrastructural criteria (see (23,24) for
reviews). The fact that ependymal cells retain some properties
of primitive glial cells is evidenced by the existence of tanycytes of the third ventricle in vertebrates. These cells span
the neuroectoderm at adult stages and are therefore presumed to
unite glial and ependymal properties in one cell type (23). Indeed, in the frog, rana pipiens, ependymal cells have been found
to span the cerebellar cortex even at adult stages substituting
for the classical Bergmann glia which are absent in these animals
(17). In adult lower vertebrates, the so-called ependymal glial
cells are able to form intercellular channels in the neighborhood of a lesion (7,22). These channels are reminiscent of those
described for ependymal cells at embryonic stages of lower vertebrates as well as for the primitive radial glia of higher vertebrates (37,38,39) and are presumed to function as contact guide-
posts for the outgrowing axonal processes. In the mouse, these
channels seem to originate as spaces between radially-oriented
primitive glia (39). The capacity to reform these channels in
ependymal cells during regeneration appears to be lost or suppressed in mammals. It seems, however, that mature ependyma in
adult higher vertebrates express at least one property which is
reminiscent of a more primitive potential, the expression of C1
antigen.

REFERENCES

(1) Bignami, A.; Eng, L.F.; Dahl, D.; and Uyeda, C.T.   1972.
     Localization of the glial fibrillary acidic protein in
     astrocytes by immunofluorescence. Brain Res. 43: 429-435.

(2) Bunge, R.P. 1968. Glial cells and the central myelin
     sheath. Physiol. Rev. 48: 197-251.

(3) Caviness, V.S., Jr., and Rakic, P. 1978. Mechanisms of
     cortical development: A view from mutations in mice. Ann.
     Rev. Neurosci. 1: 297-326.

(4) Dahl, D. 1976. Glial fibrillary acidic protein from bovine
     rat brain. Degradation in tissues and homogenates. Biochim.
     Biophys. Acta 420: 142-145.

(5) Dahl, D., and Bignami, A. 1973. Immunochemical and immuno-
     fluorescence studies of the glial fibrillary acidic protein
     in vertebrates. Brain Res. 61: 279-293.

(6) Detwiler, S.R., and Kehoe, K. 1939. Further observations
     on the origin of the sheath cells of Schwann. J. Exp. Zool.
     81: 415-435.

(7) Egar, M., and Singer, M. 1972. The role of ependyma in
     spinal cord regeneration in the urodele, Triturus. J. Exp.
     Neurol. 37/2: 422-430.

(8) Eng, L.F.; Vanderhaegen, J.J.; Bignami, A.; and Gerstl, B.
     1971. An acidic protein isolated from fibrous astrocytes.
     Brain Res. 28: 351-354.

(9) Franke, W.W.; Schmid, E.; Osborn, M.; Weber, K. 1978. Dif-
     ferent intermediate-sized filaments distinguished by immuno-
     fluorescence microscopy. Proc. Nat. Acad. Sci. USA 75: 5034-
     5038.

(10) Harrison, R.G. 1924. Neuroblast versus sheath cell in the
     development of peripheral nerves. J. Comp. Neurol. 37: 123-
     205.

(11) Hilber, H. 1943. Experimentelle Studien zum Schicksal des
     Rumpfganglienleistenmaterials. Arch. Entw.-Mech. Organ.
     142: 100-120.

(12) His, W. 1889. Neuroblasten und deren Entstehung im embry-
     onalen Mark. Abh. Kgl. Sachs. Ges. Wiss. Math. Phys. Kl. 15:
     313-372.

(13) Huntington, H.W., and Terry, R.D. 1966. The origin of the
     reactive cells in cerebral stab wounds. J. Neuropathol. Exp.
     Neurol. 25: 646-653.

(14) Imamoto, K.; Paterson, J.A.; and Leblond, C.P. 1978. Radio-
     autographic investigation of gliogenesis in the corpus cal-
     losum of young rats. I. Sequential changes in oligodendro-
     cytes. J. Comp. Neurol. 180: 115-138.

(15) Köhler, G., and Milstein, C. 1975. Continuous cultures of
     fused cells secreting antibody of predefined specificity.
     Nature 256: 495-497.

(16) Köhler, G., and Milstein, C. 1976. Derivation of specific
     antibody-producing tissue culture and tumor lines by cell
     fusion. Eur. J. Immunol. 6: 511-519.

(17) Korte, G.E., and Rosenbluth, J. 1980. Membrane specializa-
     tions in frog ependymal astrocytes. 10th Annual Meeting,
     Society for Neuroscience, Abstract 247.16.

(18) Lagenaur, C.; Sommer, I.; and Schachner, M. 1980. Sub-
     class of astroglia recognized in mouse cerebellum by mono-
     clonal antibody. Devel. Biol. 79: 367-378.

(19) Maxwell, D.S., and Krüger, L. 1965. The fine structure of
     astrocytes in the cerebral cortex and their response to fo-
     cal injury produced by heavy ionizing particles. J. Cell
     Biol. 25: 141-157.

(20) Mori, S., and Leblond, C.P. 1969. Electron microscopic
     features and proliferation of astrocytes in the corpus cal-
     losum of rats. J. Comp. Neurol. 137: 197-226.

(21) Mori, S., and Leblond, C.P. 1970. Electron microscopic
     identification of three classes of oligodendrocytes and
     preliminary study of their proliferative activity in the
     corpus callosum of young rats. J. Comp. Neurol. 139: 1-30.

(22) Nordlander, R.H., and Singer, M. 1978. The role of ependyma
     in regeneration of the spinal cord in the urodele amphibian
     tail. J. Comp. Neur. 180/2: 349-374.

(23) Oksche, A., ed. 1980. Neuroglia I. Berlin, Heidelberg,
     New York: Springer Verlag.

(24) Peters, A.; Palay, S.L.; and Webster, H. Def. 1976. The
     Fine Structure of the Nervous System. The Neurons and Sup-
     porting Cells. Philadelphia, London, Toronto: W.B. Saunders.

(25) Privat, A. 1975. Postnatal gliogenesis in the mammalian
     brain. Int. Rev. Cytol. 40: 281-323.

(26) Rakic, P. 1971. Neuron-glia relationship during granule
     cell migration in cerebellar cortex. A Golgi and electron
     microscopic study in Macacus rhesus. J. Comp. Neurol. 141:
     283-312.

(27) Rakic, P., and Sidman, R.L. 1973. Weaver mutant mouse
     cerebellum: Defective neuronal migration secondary to spe-
     cific abnormality of Bergmann glia. Proc. Nat. Acad. Sci.
     USA 70: 240-244.

(28) Ramon y Cajal, S. 1928. Degeneration and Regeneration of
the Nervous System (R.M. May, trans.) New York: Hafner, 1959.

(29) Ramón-Moliner, E. 1958. A study of neuroglia: The problem of transitional forms. J. Comp. Neurol. 110: 157-171.

(30) Rioux, F.; Derbin, C.; Marqules, S.; Joubert, R.; and Bis-
conte, J.-C. 1980. Kinetics of oligodendrocyte-like cells
in primary culture of mouse embryonic brain. Devel. Biol.
76: 87-99.

(31) Schachner, M.; Kim, S.K.; and Zehnle, R. 1981. Developmental expression in central and peripheral nervous system
of oligodendrocyte cell surface antigens (O antigens) recognized by monoclonal antibodies. Devel. Biol., in press.

(32) Schaper, A. 1897. Die frühesten Differenzierungsvorgänge
im Centralnervensystem. Arch. Entw.-Mech. Organ. 5: 81-132.

(33) Schaper, A. 1897. The earliest differentiation in the central nervous system of vertebrates. Science 5: 430-431.

(34) Schmechel, D.E., and Rakic, P. 1979. A Golgi study of radial glial cells in developing monkey telencephalon: morphogenesis and transformation astrocytes. Anat. Embryol. 156:
115-152.

(35) Schnitzer, J.; Franke, W.W.; and Schachner, M. 1981.
Immunocytochemical demonstration of vimentin in astrocytes
and ependymal cells of developing and adult mouse nervous
system. J. Cell Biol., in press.

(36) Sidman, R.L., and Rakic, P. 1973. Neuronal migration with
special reference to developing human brain: A review. Brain
Res. 62: 1-35.

(37) Silver, J., and Robb, R.M. 1979. Studies on the development
of the eye cup and optic nerve in normal mice and mutants
with congenital optic nerve aplasia. Dev. Biol. 68: 175-190.

(38) Silver, J., and Robb, R.M. 1978. Malformation of the eye
cup and optic nerve in mice with inherited ocular retarda-
tion. Invest. Ophthalmol. Vis. Sci. (Suppl.) 17: 149.

(39) Silver, J., and Sidman, R.L. 1980. A mechanism for the
guidance and topographic patterning of retinal ganglion
cell axons. J. Comp. Neur. 185: 1-33.

(40) Smart, I., and Leblond, C.P. 1961. Evidence for division
and transformation of neuroglia cells in the mouse brain
as derived from radioautography after injection of [3]H-
thymidine. J. Comp. Neurol. 116: 349-367.

(41) Sommer, I., and Schachner, M. 1981. Monoclonal antibodies
(O1 to O4) to oligodendrocyte cell surfaces: an immunocyto-
logical study in the central nervous system. Devel. Biol.,
in press.

336   M. Schachner et al.

(42) Vaughn, J.E.   1969.   An electron microscopic analysis of
     gliogenesis in rat optic nerves. Z. Zellforsch. Mikrosk.
     Anat. Abt. Histochem. 94: 293-324.

(43) Vaughn, J.E.; Hinds, P.L.; and Skoff, R.P.   1970.   Electron
     microscopic studies of Wallerian degeneration in rat optic
     nerves. I. The multi-potential glia. J. Comp. Neurol. 140:
     175-206.

(44) Vaughn, J.E., and Peters, A.   1968.   A third neuroglial
     cell type: an electron microscopic study. J. Comp. Neurol.
     133: 175-206.

(45) Vaughn, J.E., and Peters, A.   1971.   The morphology and
     development of neuroglial cells. In Cellular Aspects of
     Neural Growth and Differentiation, ed. D.C. Pease, pp.
     103-104. Los Angeles: University of California Press.

(46) Weston, J.A.   1963.   A radioautographic analysis of the
     migration and localization of trunk neural crest cells
     in the chick. Devel. Biol. 6: 279-310.

Group on Repair

Standing, left to right: Ian McDonald, Bob Barchi, Dick Quarles,
Hugh Bostock.
Seated: Itzchak Parnas, Melitta Schachner, Emanuel Stadlan,
Lloyd Greene.

Neuronal-glial Cell Interrelationships, ed. T.A. Sears, pp. 339-362.
Dahlem Konferenzen 1982. Berlin, Heidelberg, New York: Springer-Verlag.

Repair
State of the Art Report

R. L. Barchi, Rapporteur
H. Bostock, L. A. Greene, W. I. McDonald,
I. Parnas, R. H. Quarles, M. Schachner, E. M. Stadlan

INTRODUCTION

The concept of repair processes in multiple sclerosis, especially
those which relate to neuronal-glial cell interactions, can be
considered on several different levels. Eventually it may be
possible to document an initial insult in man which leads directly or indirectly to the eventual development of the symptoms of multiple sclerosis. Repair at the most basic level
might be directed at elimination of this insult be it caused by
viral, humoral, or other factors. It is also probable that
secondary immunologic or other host-reactive mechanisms may
become established as a result of this initial process which
in themselves lead to the persistence of the disease state or
to its re-appearance or exacerbation. Repair might be conceived as involving intervention at this point in the train
of pathophysiologic events. Eventually the formation of lesions
within the substance of the central nervous system leads to the
characteristic symptomatic expressions of the disease in man.
The concept of repair might be considered to involve the restoration of normal function in the fiber tracts affected by
demyelination and axonal loss.

Very little is known about the earliest steps in the establishment of multiple sclerosis in man and little more has been
documented with certainty concerning the intermediate events
leading to the formation of the characteristic plaques in the
central nervous system. In view of our lack of primary information, the question of repair at these two levels was not pursued in our discussion, although its ultimate importance in
regard to control of the disease was clearly recognized. This
study group elected to concentrate instead on repair mechanisms
which might operate at the level of the plaque itself with an
aim towards symptomatic recovery of function in the damaged
area.

The characteristic multiple sclerosis plaque is a graded lesion.
In the central area of the plaque there is nearly complete loss
of myelin from all nerve fibers while characteristically most
of the axons themselves are spared. Acute demyelination in previously myelinated fibers will produce conduction block and
failure of impulse propagation as will be discussed below. Although demyelination without axonal damage is considered characteristic of multiple sclerosis, one invariably finds some fiber
damage and axonal loss in the most severely effected portion
at the center of the plaque. Damage becomes less marked in a
graded fashion towards the plaque periphery where partial rather
than complete myelin loss is seen and nerve fiber damage is
minimal. Thus, damage within a given plaque will not be uniform and will run the gamut from mild paranodal damage through
complete internodal demyelination and on to frank axonal degeneration. Thus, while the primary role of demyelination of central
axonal segments in the production of symptoms in multiple sclerosis was recognized, the small but significant contribution of
nerve fiber loss was also considered. In addition, the possibility of alternative mechanisms of damage involving distal
axonal components and synaptic function was also addressed and
potential repair mechanisms at these levels considered. Furthermore, the presence of evanescent clinical symptoms which
develop and resolve much too quickly to be attributed to

demyelination were also recognized. The possible role of edema and other transient factors in the production of such symptoms was considered.

Given the observation that multiple sclerosis plaques contain little or no myelin in their central regions, the question of repair at this level can be subdivided into three general categories: 1) that dealing with the restoration of functional myelin sheaths to the denuded axons, 2) that concerning the re-establishment of useful conduction within persistently demyelinated segments of these central axons, and 3) that addressing the possible re-establishment of functional capacity to an involved pathway or tract by utilization of alternate fibers bypassing the site of the lesion. Each of these areas will be considered in turn below.

REMYELINATION IN THE MULTIPLE SCLEROSIS PLAQUE
Morphologic examination of the MS plaque indicates three general areas: 1) a central zone in which all myelin is lost and in which there appear to be few, if any, surviving oligodendrocytes by light and electron microscopic criteria; 2) a variable, narrow surrounding ring demonstrating partial demyelination; and finally 3) the border of neighboring healthy tissue with its normal complement of oligodendrocytes and myelinated fibers. The central portion of an established plaque also exhibits marked astrocytic proliferation leading to the gross "scarring" usually apparent in the sectioned brain at autopsy which led to the initial descriptive name for this disorder. Remyelination in these areas of extensive myelin loss has been observed only rarely; occasionally fibers can be identified in which there is evidence of newly formed myelin internodal segments, but this is clearly the exception rather than the rule (27). Most often axons in a plaque remain demyelinated indefinitely.

Various reasons for the failure of these axons to acquire new myelin can be entertained. The oligodendrocytes formerly

providing the myelin in this region may have been destroyed
during its formation and myelin competent cells may not be present in the area. Oligodendrocytes capable of forming and maintaining myelin, or precursors of such cells, may be too far
away from the primary site of damage to respond to the need
for new myelin. These competent oligodendrocytes may in fact
be unable to divide in the adult to provide the new cells which
are necessary or may be inhibited from dividing by a factor or
factors elaborated within the lesion. Competent oligodendrocytes capable of division may be unable to migrate into the
lesion either due to the absence of trophic factors or to
mechanical obstruction from other components such as the astrocytes populating that area. Finally, the axons themselves may
in fact be damaged in such a way that either appropriate trophic
factors are no longer produced or their surface membranes no
longer provide a suitable substrate for myelination even if
presented to the appropriate competent cell. Development of
techniques to stimulate effective repair of myelin in the plaque
demands some understanding of each of these possibilities.

Oligodendrocytes in the Multiple Sclerosis Plaque
The question of the presence or absence of viable oligodendrocytes within the plaque must first be answered definitively.
Although classical morphologic techniques suggest that most
oligodendrocytes have in fact been lost, these techniques depend on recognition of characteristics of the normal architecture of this cell type. It is possible that the disease process
produces significant changes in cellular morphology and that
oligodendrocytes do remain which are not recognized as such
using these standard morphological parameters. Application of
appropriate immunocytochemical techniques to these plaques which
would allow recognition of surface or cytoplasmic antigens
specific for the oligodendroglial cell should provide a definitive answer to this central question (see Schachner et al., this
volume).

Several experimental observations in addition to the standard
light and EM studies referred to above provide support for the

concept that oligodendroglia are no longer present within the plaque. Whether they are the primary site of attack or whether the myelin is locally destroyed (or whether both processes are involved) remains uncertain. There is recent evidence (26) for focal active destruction by cellular elements which may be different in nature from that seen in experimental allergic encephalomyelitis, a process in which an immune-mediated event is induced directly against the compact myelin or one of its components, or in other experimental models of demyelination. On the other hand, immunocytochemical examination of MS lesions showed an early loss of staining for the myelin-associated glycoprotein (MAG) in the periphery of developing plaques where staining for myelin basic protein and the histological appearance of Luxol fast blue stained myelin appeared normal (19). Since MAG, produced by the oligodendrocyte, is concentrated in the most distal part of the myelin sheath closest to the axon and farthest from the parent oligodendrocyte cell body (19,35), this observation suggests an initial insult to the oligodendrocyte which affects metabolic support of the sheath rather than an attack on the structure of the compact myelin from without. Thinning of the myelin sheath during early demyelination may represent a "dying back" process with retraction of membrane elements toward the failing oligodendrocyte cell body with consequent unrolling of the myelin sheath. Such a mechanism, however, would require dissolution of tight junctional components normally involved in the maintainance of compact myelin structure.

Further inferential support for the absence of competent oligodendrocytes in the MS plaque can be drawn from several experimental models of demyelination in the central nervous system. Local injection of lysolecithin into the CNS produces profound local demyelination (3) with resultant block of neuronal conduction (34). In this lesion remyelination is rapid and usually complete, suggesting that oligodendrocytes are spared. Demyelination can also be produced in the CNS by local introduction of diphtheria toxin (15). Once again profound demyelination and conduction block occur (25). However, in this case very

little remyelination occurs, suggesting that oligodendrocytes
within the lesion are destroyed by the toxin. The marked difference between these two demyelinating lesions, one of which
recovers with normal myelin formation by oligodendrocytes, the
other remaining persistently demyelinated, may reflect the preservation of competent oligodendrocytes within the former lesion but not in the latter. Again, unequivocal confirmation
of the absence of oligodendrocytes in the diphtheria lesion
might best be supported by further studies using oligodendro-
cyte-specific immunocytochemical markers.

Proliferation and Migration of Oligodendrocytes

Peripheral nerve axolemma is known to be mitogenic to Schwann cells
in culture (see Bunge, this volume). Similar studies have not been
done with oligodendrocytes. It is important to determine whether
central axonal membrane is in fact a mitogen for oligodendrocytes, whether this stimulatory activity is effective throughout the life of the oligodendrocyte or only within certain time-
frames or windows during development, and whether the demyelinated axon from MS tissue is an equally effective stimulus to
cell division.

Other factors which might stimulate oligodendrocytes proliferation or migration must be sought since their selective absence
in the MS plaque might be contributing to the failure of myelin
formation. Recent preliminary experiments (Trenkner & Herrup,
personal communication) suggest that Epidermal Growth Factor
(EGF), a well-characterized polypeptide that is mitogenic for
a variety of cell types, stimulates proliferation of oligodendrocytes in culture. Significantly, this effect is seen only
during a brief period or "window" during oligodendrocyte development, underscoring the point that these cells may respond to
various stimuli only at specific times in their life cycle.
Further work is clearly needed in an effort to identify other
relevant factors which stimulate oligodendrocyte proliferation
and to define the temporal sequence of their activity.

Factors that modify the duration of sensitivity of oligodendrocytes to known mitogens will also require study. Both questions
might best be explored in tissue culture systems.

Additional thought must be given to other situations in addition to the absence of trophic components which might result
in the inhibition of proliferation and/or migration of competent oligodendrocytes or their precursors. Enzymes released
by macrophages and similar cells within the lesion may destroy
soluble protein factors or modify membrane surface receptors
in such a way as to inhibit their effect. Studies of the effects of protease inhibitors or repair processes in demyelinating lesions might therefore be profitable. Similar destructive
effects could arise from the action of immune system components.

If the destruction of oligodendrocytes within the MS plaque is
a critical element in the prevention of new myelin formation,
we must look further to the remaining normal oligodendrocytes
at the periphery of the lesion. What factors prevent these
cells from moving into the plaque and providing new myelin much
as adult Schwann cells do in acutely denervated peripheral
nerve? One possibility for this failure is an inability of mature oligodendrocytes to divide and/or migrate in response to
injury or to an appropriate mitogen or trophic factor. Evidence
regarding this point remains controversial; several model systems can be reviewed which might shed some light on this issue,
although in each case major questions of interpretation or
extrapolation to the adult MS lesion can be raised. CNS demyelination produced by CSF barbotage is followed by remyelination
(8,9). It has been suggested that oligodendrocytes within the
lesion are lost during the phase of demyelination, but this
question has not been examined using modern techniques. Examination of remyelination in the distal stump of transected tadpole optic nerve suggests that radial astrocyte processes form
a framework for migrating oligodendrocytes which subsequently
remyelinate the regenerating neurites (31). Here the major

reservation is the relatively immature nature of the tissue involved. Studies presently in progress using tissue culture by
Bunge et al. have provided an opportunity to study the interaction of oligodendrocytes or Schwann cells with unmyelinated
axons from dorsal root ganglion explants. These studies demonstrate that explanted oligodendrocytes added to such a neural
culture will in fact migrate to and associate with axons, subsequently producing myelin. Their ability to migrate from the
point of addition and the extent of myelin formed are both
limited, however. Again the major difficulty in extrapolating
these observations to the MS lesion is the experimental use of
immature oligodendrocytes from seven-day-old animals. Similar
experiments with adult oligodendrocytes have not yet been done.
Astroglial cells were present in these cultures; their positive
or negative influence on oligodendrocyte migration and myelin
formation has not been defined.

Although there is evidence suggestive of cell division in the
mouse lysolecithin lesion during remyelination (17), it has not
been definitively shown that oligodendroglial cells in adult
human brain are capable of division and extensive migration
within a demyelinating lesion. Clearly this is a major question which remains to be answered. Several possible approaches
to providing such an answer in model systems were suggested in
discussion. Thymidine incorporation studies should be undertaken in the diphtheria and lysolecithin models of demyelination
along with immunocytochemical labelling for definitive identification of glial elements. Definitive evidence for recent
cell division and for migration of cells within these lesions
should be sought. Possibly the most direct answer may come
from culture studies in which immature oligodendrocytes are re-
placed by mature cells from adult donors.

It is possible that immature or precursor glial cell elements
are present in adult brain which could give rise to myelin-
competent oligodendrocytes but which at present would not be
identified by routine morphological techniques. Such a situation

might be destroyed by thymidine incorporation studies in adult
demyelinating lesions. Oligodendrocyte markers such as O3 and
O4 antigens might be useful in identifying such cells (see Schachner
et al., this volume). The development of immunocytochemical markers for progressively more immature oligodendrocytes would be
most helpful in this regard and could profitably be used in association with the above experiments to identify such precursors.

The ability of demyelinated axons to support myelin formation
has been demonstrated in several systems. Peripheral Schwann
cells introduced into areas of experimentally produced CNS demyelination will form normal myelin around axons in the immediate vicinity (2,13). Uncharacteristically, however, these
cells will not migrate outward to ensheath all of the exposed
axons available. Similarly, MS plaques in the spinal cord can
occasionally be found in which Schwann cells from adjacent root
entry zones have invaded and remyelinated a portion of these
fibers (11,20). Again, however, the areas of remyelination are
quite restricted. Such observations suggest that the axonal
signal triggering the actual initiation of myelin formation
is still present, but suggest the existence of some impediment
to migration of the myelinating cell within the demyelinated
region.

The feasibility of directly introducing oligodendrocytes or
Schwann cells into the CNS in order to remyelinate MS plaques
was considered. Experiments of this type are presently underway in model systems (Bunge and Aguayo, personal communication).
A number of practical difficulties were raised which presently
contraindicate clinical application of this approach. Among
these was the consideration that it would be extremely difficult to precisely localize all the symptomatic plaques requiring treatment.

Astrocytes and Oligodendrocyte Migration

The MS plaque is characterized by extensive astrocytic proliferation. The presence of these astrocytes could theoretically

constitute either an aid or an impediment to oligodendrocyte movement and myelination. A positive role for astrocytes during development involving guidance of cellular migration has been suggested by several studies involving normal CNS elements (e.g., radial
glial cells, Bergmann glia) and CNS mutants (e.g., Weaver mutant)
(see Rakic, this volume). Alternatively, a dense tangle of reactive
astrocytes within a plaque could present a mechanical barrier
to or inadequate substrate for subsequent cell movement. Cytoplasmic or surface components of astrocytes mediating some of
these "positive" functions may not be expressed on the surface
of reactive astrocytes present in the MS plaque. For example,
the cytoplasmic antigen C1 can be identified in radial glial
fibers at very early stages of development and persists in
Bergmann glia and Muller cells (see Schachner et al., this volume).
A second antigen, M1, appears later in white matter astrocytes
but not in those containing C1. Preliminary experiments suggest that if the brain is damaged by a stab wound, the reactive
astrocytes in the vicinity of the wound are induced to express
M1 and not C1 antigen.

Astrocytes may vary in their effects on oligodendrocyte migration; this area requires further study both in vivo and in tissue culture systems using astrocytes of known type as identified by multiple immunochemical markers. Similarly, positive
and negative aspects of astroglial proliferation in a demyelinated area can be proposed for the electrophysiologic function
of the axon (see below). The timescale of development of astro-
cytosis in a newly formed MS plaque is not known and therefore
should be studied.

RESTORATION OF FUNCTIONAL CONDUCTION IN PERSISTENTLY DEMYELINATED FIBERS

Ample clinical evidence exists to support the concept that
functional conduction can be re-established through a demyelinated MS plaque without remyelination. The paucity of remyelination in pathologically examined plaques taken in con-
junction with the fundamental relapsing and remitting nature

of the disease itself suggests that this should be so, and recent clinico-pathologic correlations of a severely demyelinated
optic nerve (see Bostock and McDonald, this volume) provide
morphological verification. If this is so, the conduction
block produced in myelinated axons by the removal of internodal myelin must be overcome by some alternative mechanism.

The nature of the conduction block in demyelination should first
be reviewed briefly. In saltatory conduction, current passing
inward at a given node during an action potential spreads along
the adjacent internode. Most current moves longitudinally rather
than radially due to the high radial resistance and low radial
capacitance provided by the myelin sheath. The next nodal region provides a relatively low resistance pathway for outward
current flow, and much of the current exits here to complete the
circuit through the extracellular fluid. This outward nodal
current flow depolarizes the membrane resulting in the generation
of an action potential. This process may then be repeated with
the impulse effectively jumping from node to node. The internodal length, myelin thickness, etc., are all optimized by unknown signals between axon and oligodendroglia to provide maximal passive spread of currents along the internode with minimal
losses.

When the myelin of an internode or even a significant portion
of the myelin immediately adjacent to the nodal membrane is
removed or lifted from the axonal membrane, the passive properties of the axon change dramatically. Current flowing down the
last intact internode from the previous node encounters a large
area of membrane having a low resistance and high capacitance.
Because of the extra membrane capacity, this area of membrane
takes much longer than normal to be depolarized to threshold,
and internodal conduction times have been observed to increase
as much as 25-fold in experimental demyelination (30). The
duration of the depolarizing current is limited by the driving
action potential, and at low rates of depolarization the threshold is increased, so that more than a limited loss of paranodal

myelin results in conduction failure (22). Even when the extent of demyelination is insufficient to block single impulses,
conduction of trains of impulses is impaired (30) probably because the prolonged current flow causes accumulation of intracellular sodium and extracellular potassium ions. The difficulty
is compounded by the observation that sodium channel density,
while very high at the node, is very low in the internodal membrane under the myelin, and that, in addition, voltage-dependent
potassium channels are present in this membrane.

Conduction in Demyelinated Internodes

Recently, Bostock and Sears (5,6) have been able to demonstrate
in the peripheral nervous system the restoration of conduction
along such demyelinated zones which has previously represented
sights of conduction failure. Conduction along these demyelinated zones appeared first at about six days after demyelination by diphtheria toxin, although evidence of unstable
conduction could be seen as early as four days after treatment.
In nerves demyelinated with diphtheria toxin, conduction along
the demyelinated axon appeared continuous at the resolution
of the electrode system used (about 250µ), while in lysolecithin treated nerves the demyelinated segments developed
discrete "hot spots" of inward current suggestive of local clustering of active ionic channels (4).

The appearance of continuous conduction in these fibers has been
attributed to the appearance of sodium channels in the internodal
membrane at a density not present immediately after demyelination.
It is not known whether these channels represent newly synthesized protein or are the result of lateral movement of channels
formerly present in high density in the nodal membrane. Experimental current recordings indicate that the high density nodal
concentration of channels has become much less prominent after
demyelination at a time when single impulse conduction is restored. This might indicate removal of channels from the membrane or lateral spread within the plane of the membrane.

At this time it is not known whether the passive properties of
the axonal membrane in the region of demyelination change during
the time at which conduction is restored. Any increase in membrane resistance would favor effective depolarization; this could
result from alteration in membrane leakage currents or from close
apposition of glial elements to the axonal surface. The diameter
of the demyelinated segment does decrease (28). This certainly
will favor re-establishment of conduction by decreasing membrane
surface area and increasing longitudinal resistance, both factors tending to increase outward current density per unit area
of membrane in the proximal demyelinated segment.

Simple restoration of conduction for single impulses through
such a demyelinated zone does not ensure normal functioning
of the pathway in which these axons are involved. Information
in many neural pathways is contained not only in the presence
or absence of impulses, but also in their frequency and pattern.
Thus the ability of an axon to convey information accurately is
in part a reflection of its capacity to transmit trains of impulses at the optimum frequency usually encountered in that
pathway. Restoration of continuous conduction in demyelinated
segments is usually associated with limited high frequency response capability; the appearance of a train of impulses in
rapid succession will usually result in block, with failure to
transmit faithfully some of the impulses in the train. Thus
even after restoration of continuous conduction, function in
pathways requiring preservation of higher frequency information
will be abnormal. This fact may account in part for the persistent impairment of vibratory sensation in a large percentage
of MS patients. Furthermore, loss of myelin near branch points
of axons near synaptic terminals may result in persistent or
intermittent failure of conduction along a branch, variable
synaptic function, or reentrant excitation phenomena even after
establishment of continuous conduction.

External Agents and Conduction

Attempts have been made to improve the reliability of transmission across a demyelinated region in which sodium channel density

has increased sufficiently to support conduction. It can be
predicted that factors which prolong the action potential at
the last normal node, and therefore increase the duration of
inward current flowing longitudinally toward the demyelinated
region, will enhance the safety factor of transmission in
conducting fibers, or allow transmission to occur in marginal
ones (32). Several methods of prolonging the action potential
have been pursued.

The simplest way to achieve this prolongation is to reduce am-
bient temperature. It can be shown in the diphtheria toxin demyelination model that variation in temperature up or down by
a small amount will interconvert an axon from a conducting to
a non-conducting state. Decreasing temperature by as little
as 0.5° C will restore conduction in an otherwise blocked axon
(29). This observation correlates well with the clinical observation of symptomatic worsening of MS patients in hot weather,
under hot showers, etc., and the anecdotal reports of improvement in function on exposure to cold.

Alternatively, pharmacologic agents are available which have
been shown to prolong the duration of the action potential.
Chemicals which block voltage-dependent potassium channels
such as 4-aminopyridine (4-AP) or tetraethylammonium ion (TEA)
will retard the repolarization process at nodes containing these
channels, and therefore prolong the duration of the depolarized
state. 4-AP significantly improves the competency of conduction in demyelinated regions of the diphtheria toxin model (33).
Its clinical usefulness, however, appears to be limited by other
effects of this agent, especially on synaptic transmission. In
a limited clinical test the intervention of cortical seizure
activity prevented further administration (Bostock et al., unpublished observations). Its future usefulness appears limited.

Another method of prolonging the nodal action potential is to
delay the voltage-dependent inactivation of the sodium channel.
This effect can be produced by a class of polypeptide neurotoxins

elaborated by several species of scorpion and certain sea anemones. Scorpion toxin will improve the reliability of conduction through a demyelinated segment in an experimental system
(7). Its usefulness in humans is doubtful if for no other
reason than the obvious problems of delivery of such a peptide
of 6000-8000 MW across the blood-brain barrier.

Research is needed to identify smaller molecules capable of
entering the CNS which would function in a manner similar to the
scorpion toxins or 4-AP without producing the unacceptable side
affects inherent in these agents. One fruitful direction for
research might be to search for modifiers with special affinity
for the regions of demyelination so that unwanted effects in
areas of normal CNS matter can be minimized.

A final pharmacologic approach to improvement of conduction in
demyelinated segments would be to lower the threshold for action potential generation such that a smaller depolarization
is required for conduction to be initiated. One method of ac-
complishing this goal is to alter membrane surface charge by
reducing ionized calcium concentration. Hypocalcemia has been
demonstrated to produce transient improvement in MS patients
(10). A major drawback to the pursuit of pharmacologic agents
having comparable effects is the enhancement of the unpleasant
positive symptoms of MS due to hyperirritability of some damaged
fibers and the possibility of inducing uncontrolled discharges
in other pathways.

Plastic and Functional Recovery
Not all recovery of function following focal demyelination need
involve restoration of conduction through the affected area.
The plasticity and redundancy of certain pathways in the CNS
may allow alternative unaffected components of the nervous system to take over functions previously mediated by the damaged
pathway. Undoubtedly, plasticity contributes significantly to
maintenance of function following lesions in many pathways.
However, persistant deficits in MS may indicate that this reserve

has already been compromised by multiple lesions spread through
the CNS. Although augmentation of plastic repair in the CNS
may provide some potential benefit, direct experimental approach
is difficult.

Functional Recovery from Very Brief Symptomatic Episode

Symptomatic episodes occur in MS the duration of which are too
brief to be ascribed to demyelination of the kind associated
with typical plaque formation. Various mechanisms for such
episodes have been proposed, including the effects of local
edema on conduction and partial paranodal myelin damage. Functional recovery from such symptoms in MS may involve features
not considered at all in previous discussion. Further study
of these transient phenomena and their potential causes is
needed.

MODEL SYSTEMS FOR THE STUDY OF REPAIR IN MS

Model systems applicable to various aspects of potential repair
mechanisms in MS have been cited in earlier sections of this
report. They will be briefly consolidated here along with
several others of potential value. The list is not intended
to be inclusive, but only representative of the experimental
systems available.

In addition to traditional morphological procedures, a useful
method for evaluating the similarity of lesions in model systems to multiple sclerosis plaques will be immunocytochemical
examination. Staining of paraffin sections of multiple sclerosis tissue with a limited number of antisera has already demonstrated similarities and differences between multiple sclerosis
lesions and those in other demyelinating conditions (18,19,21,36).
As antisera to additional myelin and oligodendroglial components
become available, more detailed molecular maps of demyelinating
lesions will be possible.

Models of CNS Demyelination - Physically- or Chemically-Induced

Injection of lysolecithin into the CNS produces an acute demyelinating lesion characterized by successful remyelination and

recovery of function (3). In this model, oligodendroglial cell
bodies appear to be spared. Similar local CNS injection of
diphtheria toxin also produces acute demyelination (15). Here
oligodendrocyte remyelination is minimal or absent, suggesting
that oligodendroglia within the lesion have been destroyed (16).
Comparison in detail of these two models using current immunocytochemical and radiolabelling techniques may prove valuable.
Both provide methods of producing demyelinated fibers which
may serve as a substrate for implantation of exogenous oligodendrocytes, Schwann cells, trophic factors, etc. Since remyelination is virtually absent in the diphtheria lesion, factors affecting the rate and extent of astrocyte proliferation
might best be studied here. Cuprizone intoxication in mice
produces continuing demyelination which recovers when drug administration is stopped (1,23).

CSF barbotage (8) and acute compression (14) may provide a
useful model of acute demyelination. Lesions are superficial;
oligodendrocytes may be damaged, but remyelination does occur.

Models of Viral-Induced Demyelination
Virus-induced demyelination has been covered more extensively
in other sections of this workshop and several models will be mentioned here only as examples (see Fields and Weiner, this volume).

The mouse coronavirus JHM will produce acute demyelinating
lesions of the mouse brain and spinal cord six to eight days
after infection. This virus appears selective for oligodendrocytes. These cells die, but remyelination does appear to
occur (17). This model may provide insight into mechanisms
of oligodendrocyte proliferation and migration under certain
pathologic conditions. Canine distemper virus causes severe
demyelination in dogs, but little is known about the cellular
pathology. Transmission to other species has not been accomplished. Theiler's virus produces a secondary inflammatory demyelinating syndrome in mice after first infecting anterior
horn cells. A form of progressive multifocal leukoencephalopathy has been produced in immunosuppressed monkeys. In this

disorder extensive demyelination occurs with preservation of
axons. Withdrawal of immunosuppressants might lead to secondary
inflammatory response which could even more closely mimic MS.
The model has proven very hard to produce consistently.

Visna, a CNS viral infection of sheep, has many similarities
to MS. After innoculation there is a variable incubation period
of months to years. Relapses and remissions in symptoms are
seen, and pathological analysis reveals multifocal demyelinating
lesions and gliosis. Unfortunately the disease can only be
produced in certain strains of sheep and does not appear to
adapt to smaller mammals or primates.

Models Induced by Immunization with Nervous Tissue Elements

The syndrome of experimental allergic encephalomyelitis can
be produced in a variety of species by immunization with crude
CNS, purified myelin, or certain myelin protein components.
These models resemble MS in some aspects and differ in others.
They are discussed in detail by Arnason and Waksman (this volume)
and will not be considered further here.

Models for Studying Neuronal-glial Interactions In Vitro

Various tissue culture systems appear to hold the most promise
for directly relevant experiments concerning astrocyte and oligodendrocyte proliferation, migration, the interactions between
astrocytes and oligodendrocytes, and the interactions between
both cell types and neurons. Pure cultures of astrocytes or
oligodendroglial cells should provide a system to screen for
and subsequently study mitogens and trophic factors which influence cell division and migration respectively. Subsequently
studies with mixed cultures should shed light on the mutual inhibitory or facilitatory interactions between these cell types.
Specific attention must be directed to temporal compartmenta-
tion of certain responses during various stages of cell development. Finally, mixed cultures with neuronal elements capable
of being myelinated, such as the dorsal root ganglion preparation of Bunge, could provide the ultimate in vitro testing

ground for oligodendrocytes in terms of competency for myelin
formation during various stages of development, as well as factors promoting or inhibiting migration to and interaction with
axons. Addition of astrocytes to the system under controlled
conditions will further help clarify this dimension of neuronal-
glial interaction.

Electrophysiological Models

At present it seems reasonable to assume that observations made
with diphtheria toxin demyelination in the PNS will be directly
relevant to conduction in the CNS, since to date all the physiologic properties of normal and demyelinated PNS fibers have
found their counterpart in CNS axons when properly studied (24).

Further work with the diphtheria toxin and lysolecithin induced
demyelination models in rat ventral roots will undoubtedly be
fruitful. Newer models of acute demyelination using anti-
galactocerebroside antiserum may provide additional information.
More detailed analysis of these systems at the single fiber
level using voltage clamp techniques may soon be possible.

RELEVANCE OF REPAIR MECHANISMS TO MULTIPLE SCLEROSIS

This report has discussed a variety of potential repair processes which could come into play in the demyelinated MS plaque.
An appropriate summation might put these into perspective in
regard to their probability of occurring naturally in the human
disease state, and their relative potential for future manipulation in the treatment of MS.

Although remyelination of denuded axons is functionally the most
satisfying repair mechanism possible (34), it appears to be a
very minor feature in the natural history of MS plaques. Re-
establishment of conduction through demyelinated segments of
axon, on the other hand, may play a major role in the recovery
of function after an acute exacerbation in MS. Synaptic plasticity and remodeling of alternate pathways may play an important role whose extent is as yet unappreciated.

On the other hand, one can consider these broad categories of
repair mechanisms in terms of their potential for exploitation
and manipulation for more effective treatment in MS. We know
very little about the control of plasticity in the human CNS,
and it seems unlikely that we will be able to exploit this as
a repair process in the foreseeable future any more effectively
than is being done by the brain itself. Restoration of conduction in demyelinated segments has immediate applicability and
also the potential for pharmacologic manipulation. The possibility of significant advances in this area seem very real.
On the other hand, most pharmacologic approaches carry the
probability of complications from unwanted effects in other
neuronal pathways. Furthermore, problems such as the fidelity
of high frequency response, discussed above, may well place a
natural limit on the extent of recovery expected from this
approach in the severely disabled MS patient. Induction of remyelination, on the other hand, has **major** potential for a high degree
of physiologic repair with maximal recovery of function and on
scientific grounds seems to have also a real potential for
eventual manipulation.

Ultimately the control of MS will result from an understanding
of the underlying primary etiologic event, and the most effective form of "repair" will be prevention of the initial insult
or blockage of its further progression. There is no question,
however, that work toward repair of damage at the site of the
lesion must progress in parallel, and the probability exists
that such work will yield clinically useful information in the
near future.

REFERENCES

(1) Blakemore, W.F. 1973. Remyelination of the superior
     cerebellular peduncle in the mouse following demyelina-
     tion induced by feeding cuprizone. J. Neurol. Sci. 20:
     73-83.

(2) Blakemore, W.F. 1973. Remyelination of CNS axons by
     Schwann cells transplanted from the sciatic nerve. Nature
     226: 68-69.

(3) Blakemore, W.F.; Eames, R.A.; Smith, K.J.; and McDonald,
     W.I. 1977. Remyelination in the spinal cord of the cat
     following intraspinal injections of lysolecithin. J.
     Neurol. Sci. 33: 31-43.

(4) Bostock, H.; Hall, S.M.; and Smith, K.J. 1980. Demyelin-
     ated axons form 'nodes' prior to remyelination. J. Physiol.
     308: 21P-23P.

(5) Bostock, H., and Sears, T.A. Continuous conduction in de-
     myelinated mammalian nerve fibers. Nature 263: 786-787.

(6) Bostock, H., and Sears, T.A. 1978. The internodal axon
     membrane: electrical excitability and continuous conduc-
     tion in segmental demyelination. J. Physiol. 280: 273-301.

(7) Bostock, H.; Sherratt, R.M.; and Sears, T.A. 1978. Over-
     coming conduction failure in demyelinated nerve fibers
     by prolonging action potentials. Nature 274: 385-387.

(8) Bunge, M.B.; Bunge, R.P.; and Ris, H. 1981. Ultrastruc-
     tural study of remyelination in an experimental lesion in
     adult cat spinal cord. J. Physiol. Biochem. Cytol. 10:
     67-94.

(9) Bunge, R.P.; Bunge, M.B.; and Ris, H. 1960. Electron
     microscopic study of demyelination in an experimentally
     induced lesion in adult cat spinal cord. J. Biophys.
     Biochem. Cytol. 7: 685-686.

(10) Davis, F.A.; Becker, R.O.; Michael, J.A.; and Sorensen,
     E. 1970. Effects of intravenous sodium bicarbonate di-
     sodium edetate (Na$_2$EDTA), and hyperventilation on visual
     and oculomotor signs in multiple sclerosis. J. Neurol.
     Neurosurg. Psychiat. 33: 723-732.

(11) Fergen, I., and Popoff, N. 1966. Regeneration of myelin
     in multiple sclerosis. Neurology 16: 364-372.

(12) Gledhill, R.F., and McDonald, W.I. 1977. Morphological
     characteristics of central demyelination and remyelination:
     a singel fibre study. Ann. Neurol. 1: 552-560.

(13) Harrison, B.M. 1980. Remyelination by cells introduced into a stable demyelinating lesion in the central nervous system. J. Neurol. Sci. 46: 63-81.

(14) Harrison, B.M., and McDonald, W.I. 1977. Remyelination after transient experimental compression of the spinal cord. Ann. Neurol. 1: 542-551.

(15) Harrison, B.M.; McDonald, W.I.; and Ochoa, J. 1972. Central demyelination produced by diphtheria toxin: an electron microscope study. J. Neurol. Sci. 17: 293-302.

(16) Harrison, B.M.; McDonald, W.I.; and Ochoa, J. 1972. Remyelination in the central diphtheria toxin lesion. J. Neurol. Sci. 17: 293-302.

(17) Herndon, R.M.; Price, D.L.; and Weiner, L.P. 1977. Regeneration of oligodendroglia during recovery from demyelinating disease. Science 195: 693-694.

(18) Itoyama, Y.; Sternberger, N.H.; Quarles, R.H.; Cohen, S.R.; and Webster, H. de F. 1979. Immunocytochemical observations on demyelinating lesions in experimental allergic encephalomyelitis (EAE). Abstracts, Soc. Neurosci. 5: 512.

(19) Itoyama, Y.; Sternberger, N.H.; Webster, H. de F.; Quarles, R.H.; Cohen, S.R.; and Richardson, E.P. 1980. Immunocytochemical observations on the distribution of myelin-associated glycoprotein and myelin basic protein in multiple sclerosis lesions. Ann. Neurol. 7: 167-177.

(20) Itoyama, Y.; Trapp, P.; McIntyre, L.; Sternberger, P.; Richardson, E.; and Webster, H. de F. 1979. Remyelination of CNS fibers by Schwann cells in multiple sclerosis: immunocytochemical observation. J. Neuropath. Exp. Neurol. 38: 323 (abstract).

(21) Itoyama, Y.; Walker, D.L.; Richardson, E.P.; Sternberger, N.H.; Padgett, P.L.; Quarles, R.H.; and Webster, H. de F. 1980. Papova virus, myelin-associated glycoprotein and myelin basic protein in progressive multifocal leukoencephalopathy. J. Neuropath. Exp. Neurol. 39: 363 (abstract).

(22) Koles, Z., and Rasminski, M. 1972. Computer simulation of conduction in demyelinated nerve fibers. 1972. J. Physiol. 227: 351-364.

(23) Ludwin, S.K. 1978. Central nervous system demyelination and remyelination in the mouse: An ultrastructural study of cuprizone toxicity. Lab. Invest. 39: 597-612.

(24) McDonald, W.I. 1974. Pathophysiology in multiple sclerosis. Brain 97: 179-196.

(25) McDonald, W.I., and Sears, T.A. 1970. The effects of experimental demyelination on conduction in the central nervous system. Brain 93: 583-598.

(26) Prineas, J.W., and Connell, F. 1978. The fine structure of chronically active multiple sclerosis plaques. Neurology (Minneap.) 28 (2): 68-75.

(27) Prineas, J.W., and Connell, F. 1979. Remyelination in multiple sclerosis. Ann. Neurol. 5: 22-31.

(28) Raine, C.S.; Wisniewski, H.; and Prineas, J. 1969. An ultrastructural study of experimental demyelination and remyelination. II. Chronic experimental allergic encephalomyelitis in the peripheral nervous system. Lab. Invest. 21: 316-327.

(29) Rasminsky, M. 1973. The effects of temperature on conduction in demyelinated single nerve fibers. Arch. Neurol. Chicago. 28: 287-292.

(30) Rasminsky, M., and Sears, T.A. 1972. Internodal conduction in undissected demyelinated fibers. J. Physiol. 227: 323-350.

(31) Reier, P.J., and Webster, H. de F. 1974. Regeneration and remyelination of Xenopus tadpole optic nerve fibers following transection or crush. J. Neurocytol. 3: 591-618.

(32) Schauf, C.L., and Davis, F.A. 1974. Impulse conduction in multiple sclerosis: A theoretical basis for modification by temperature and pharmacological agents. J. Neurol. Neurosurg. Psychiat. 37: 152-161.

(33) Sherratt, R.M.; Bostock, H.; and Sears, T.A. 1980. Effects of 4-aminopyridine on normal and demyelinated mammalian nerve fibers. Nature 283: 570-572.

(34) Smith, K.J.; Blakemore, W.F.; and McDOnald, W.I. 1979. Central remyelination restores secure conduction. Nature 280: 395-396.

(35) Stenberger, N.H.; Quarles, R.H.; Itoyama, Y.; and Webster, H. de F. 1979. Myelin-associated glycoprotein demonstrated immunocytochemically in myelin and myelin-forming cells of developing rat. Proc. Natl. Acad. Sci. 76: 1510-1514.

(36) Winchell, K.H.; Sternberger, N.H.; Quarles, R.H.; and Webster, H. de F. 1980. Myelin-associated glycoprotein and basic protein in hexachlorophene lesions. Trans. Amer. Soc. Neurochem. 11: 159 (abstract).

List of Participants

AGUAYO, A.J.
Division of Neurology
Montreal General Hospital
Montreal, Quebec H3G 1A4, Canada

Field of research: Cell inter-
actions

ARNASON, B.G.W.
Dept. of Neurology
University of Chicago
Chicago, IL 60637, USA

Field of research: Neuroimmu-
nology

BARCHI, R.L.
Dept. of Neurology
Hospital of the University
of Pennsylvania
Philadelphia, PA 19104, USA

Field of research: Molecular basis
of ion conductance mechanisms -
purification of the excitable mem-
brane sodium channel

BAUER, H.J.
Neurologische Klinik der
Universität Göttingen
3400 Göttingen, F.R. Germany

Field of research: Clinical neu-
rology, especially Multiple Sclero-
sis; immunological disorders of the
nervous system

BOSTOCK, H.
Sobell Dept. of Neurophysiology
Institute of Neurology
London WC1N 3BG, England

Field of research: Physiology of
nerve conduction, demyelination
and remyelination

BROWN, D.A.
Dept. of Pharmacology
University of London
London WC1N 1AX, England

Field of research: Glial cell function
in relation to inhibitory amino acid
action and transport

BUNGE, R.P.
Dept. of Anatomy and Neurobiology
Washington University School of Medicine
St. Louis, MO 63110, USA

Field of research: Developmental neuro-
biology

COWAN, W.M.
Salk Institute for Biological Studies
P.O. Box 85800, San Diego, CA 92138, USA

Field of research: Developmental neuro-
biology

DAVISON, A.N.
Dept. of Neurochemistry
Institute of Neurology
London WC1N 3BG, England

Field of research: Immunochemistry of
Multiple Sclerosis, biochemistry of
aging and dementia

DISTEL, H.
Institut für Medizinische Psychologie
8000 München 2, F.R. Germany

Field of research: Developmental neuro-
biology

DROZ, B.
Département de biologie
Commissariat à l'Energie Atomique
C.E.N. de Saclay, B.P. 2
91190 Gif sur Yvette, France

Field of research: Contribution of
axonal transport to maintenance of the
myelin sheath

EIGEN, M.
Abteilung Biochemische Kinetik
Max-Planck-Institut für
biophysikalische Chemie
3400 Göttingen, F.R. Germany

*Field of research: Mechanisms of
biochemical reactions (enzyme
kinetics, code reading, biopoly-
merization, sequential kinetics),
molecular self-organization:
origin and evolution of life
(theory and experiments)*

FAMBROUGH, D.M.
Dept. of Embryology
Carnegie Institute of Washington
Baltimore, MD 21210, USA

*Field of research: Plasma membrane
proteins of neurons and muscle fibers*

FELGENHAUER, K.M.A.
Neuro-psychiatrische Forschungs-
Abteilung, Universitäts-Nervenklinik
5000 Köln 41, F.R. Germany

*Field of research: Quantitation and char-
acterization of cerebrospinal fluid immunoglobulins in inflammatory diseases
of the central nervous system*

FIELDS, B.N.
Dept. of Microbiology and Molecular
Genetics, Harvard Medical School
Boston, MA 02115, USA

Field of research: Viral pathogenesis

FISCHER, G.
Institut für Neurobiologie der
Universität Heidelberg
6900 Heidelberg, F.R. Germany

*Field of research: Defined cell cul-
ture media, cell surface antigens,
cell-cell binding, cytoskeleton*

GERHARD, W.U.
Wistar Institute
Philadelphia, PA 19104, USA

*Field of research: Viral immunology,
immunopathology, mechanisms in CNS*

GREENE, L.A.
Dept. of Pharmacology
New York University Medical Center
New York, NY 10016, USA

*Field of research: Developmental
neurochemistry, mechanisms of neuronal differentiation, nerve growth
factor*

GRIFFIN, J.W.
Traylor Bldg, Rm 317
Johns Hopkins Hospital
Baltimore, MD 21205, USA

*Field of research: Axonal transport
in experimental neuropathies, axon-
Schwann cell interactions in axonal
disorders*

HILLE, B.
Physiology and Biophysics, SJ-40
University of Washington
Seattle, WA 98195, USA

*Field of research: Biophysics of excit-
able membranes, nature of ionic channels*

JOHNSON, R.T.
Dept. of Neurology
Johns Hopkins School of Medicine
709 Traylor Bldg.
Baltimore, MD 21205, USA

Field of research: Neurovirology

LIM, L.
Neurochemistry Dept.
Institute of Neurology
London WC1N 2NS, England

*Field of research: Gene expression
and its regulation in the developing
brain, hormonal involvement in differentiation of the hypothalamus*

McDONALD, W.I.
University Dept. of Clinical Neurology
Institute of Neurology
London WC1N 3BG, England

*Field of research: Clinical and ex-
perimental studies of Multiple Sclerosis, demyelination and remyelination*

MCKHANN, G.M.
Dept. of Neurology
Johns Hopkins Hospital
Baltimore, MD 21205, USA

Field of research: Neurochemistry, detection of myelin components in cerebrospinal fluid, in vitro study of oligodendroglia

MIMS, C.
Dept. of Microbiology
Guys Hospital Medical School
London SE1, England

Field of research: Viral pathogenesis, especially central nervous system

MIRSKY, R.
MRC Neuroimmunology Project
Zoology Dept., University College
London WC1E 6BT, England

Field of research: Neuroimmunology

MUGNAINI, E.
Laboratory of Neuromorphology
Dept. of Biobehavioral Sciences
Box U-154, University of Connecticut
Storrs, CT 06268, USA

Field of research: Neurocytology

NICHOLLS, J.G.
Dept. of Neurobiology
Stanford Medical School
Stanford, CA 94305, USA

Field of research: Repair and regeneration in the CNS

ORKAND, R.K.
Dept. of Physiology and Pharmacology
University of Pennsylvania
School of Dental Medicine
Philadelphia, PA 19104, USA

Field of research: Neuron-glia interactions, functional consequences of potassium accumulation in the nervous system

PAPPENHEIMER, Jr., A.M.
Biological Laboratories
Harvard University
Cambridge, MA 02138, USA

Field of research: Microbial toxins, molecular mechanisms of microbial pathogenicity

PARNAS, I.
Dept. of Neurobiology
Hebrew University
Jerusalem, Israel

Field of research: Neurobiology

QUARLES, R.H.
Section on Myelin and Brain Development
NINCDS, NIH
Park Bldg, Rm 425
Bethesda, MD 20205, USA

Field of research: Glycoproteins and proteins in myelin sheaths; their roles in glial-axonal interactions, myelinogenesis, and demyelination

RAKIC, P.
Section of Neuroanatomy
Yale School of Medicine
New Haven, CT 06510, USA

Field of research: Developmental neurobiology and neuropathology

RICKMANN, M.J.
Max-Planck-Institut für biophysikalische
Chemie, Abt. Neurobiologie
Postfach 968
3400 Göttingen, F.R. Germany

Field of research: Early stages of (astro-) glial differentiation in the neocortex

RITCHIE, J.M.
Dept. of Pharmacology
Yale School of Medicine
New Haven, CT 06510, USA

Field of research: Distribution of sodium and potassium channels in myelinated nerve

ROHRER, H.
Max-Planck-Institut für Psychiatrie
Abt. Neurochemie
8033 Martinsried, F.R. Germany

*Field of research: Neural crest
development, ciliary ganglion,
NGF receptors on dorsal root ganglion neurons during development*

SCHACHNER, M.
Institut für Neurobiologie
Universität Heidelberg
6900 Heidelberg, F.R. Germany

*Field of research: Neuroimmunology,
developmental biology of cell surface components in the nervous system*

SEARS, T.A.
Sobell Dept. of Neurophysiology
Institute of Neurology
London WC1N 3BG, England

*Field of research: Nerve conduction
in demyelination, neurophysiology of
respiratory movements*

SPITZER, N.C.
Biology Dept., B-022
University of California, San Diego
La Jolla, CA 92093, USA

*Field of research: Developmental
neurobiology*

STADLAN, E.M.
National Institute of Neurological
and Communicative Disorders and
Strokes, NIH, Federal Bldg, Rm 810
Bethesda, MD 20205, USA

*Field of research: Multiple Sclero-
sis and related disorders*

STRICHARTZ, G.R.
Dept. of Physiology, Health Science Ctr.
State University of New York
Stony Brook, NY 11794, USA

*Field of research: Structure, distri-
bution, and biosynthesis of sodium
channels in excitable cells*

TER MEULEN, V.
Institut für Virologie und Immun-
biologie, Universität Würzburg
8700 Würzburg, F.R. Germany

Field of research: Neurovirology

TREHERNE, J.E.
A.R.C. Unit, Dept. of Zoology
Cambridge, CB2 3EJ, England

*Field of research: Role of neuro-
glia in central nervous ionic
homeostasis*

WAKSMAN, B.H.
National MS Society
New York, NY 10017, USA

*Field of research: Multiple Sclero-
sis, cell-mediated immunity, suppressor mechanisms, autoimmunity*

WALICKE, P.A
Dept. of Neurology and Neurosurgery
McGill University
Montreal Neurological Institute
Montreal, Quebec H3A 2B4, Canada

*Field of research: Agents influenc-
ing transmitter choice of developing sympathetic neurons in culture, neurology*

WEBSTER, de F., H.
National Institutes of Health
Bldg 36, Rm 4D04
Bethesda, MD 20205, USA

*Field of research: Cellular mecha-
nisms of myelin formation and breakdown, immunocytochemistry of myelin
and glial cell constituents*

WEINER, H.L.
Dept. of Neuroscience G-408
Children's Hospital Medical Ctr.
Boston, MA 02115, USA

*Field of research: Mechanisms of
viral and immunological damage to
the nervous system*

Subject Index

Author Index

Dahlem Workshop Reports

Life Sciences Research Report
Editor: S. Bernhard

Volume 21

Animal Mind – Human Mind

Report of the Dahlem Workshop on
Animal Mind – Human Mind
Berlin 1981, March 22–27

Rapporteurs: M. Dawkins; W. Kintsch; H.J. Neville;
R.M. Seyfarth
Program Advisory Committee: D.R. Griffin (Chairman);
J.F. Bennett; D. Dörner; S.A. Hillyard; B.K. Hölldobler;
H.S. Markl; P.R. Marler; D. Premack
Editor: D.R. Griffin

1982. 4 photographs, 30 figures, 2 tables. X, 427 pages.
ISBN 3-540-11330-4

Background papers by D.R. Griffin, S.A. Hillyard,
F.E. Bloom, W. Hodos, J. Levy, R.H. Drent, D. Dörner,
H. Kummer, W.A. Mason, L.A. Cooper, G. Lüer,
D.J. Gillan, R.H. Kluwe, F. Klix, C.G. Beer, J.L. Gould,
C.G. Gould, C.A. Ristau, D. Robbins and
numerous specialists

The fundamental nature of animal minds is receiving
renewed attention from ethologists, neurophysiologists,
and psychologists – after many decades of neglect on the
ground that scientific investigation could not tell us
anything significant about animal thinking. But cognitive
approaches to animal behavior, and the thoughts and
feelings that many underlie it, are now the subject of
active investigation, as this Dahlem Konferenz showed.

Neuropsychological approaches (including analysis of
vent-related potentials and hemispheric specialization),
the evolutionary ecology of thinking (including riskbenefit
assessment by animals and social intelligence), internal
representation, problem solving, metacognition, and the
analysis of animal communication as evidence of thinking
were intensively reviewed at the conference.

**Springer-Verlag
Berlin
Heidelberg
New York**

This volume contains fifteen review papers together with
four thoughtful summaries of the group discussions that
allow readers to share the excitement of wrestling with
emerging ideas about new approaches to these long-standing problems.

Dahlem Workshop Reports

Life Sciences Research Report
Editor: S. Bernhard

Volume 22

Evolution and Development

Report of the Dahlem Workshop on Evolution and Development, Berlin 1981, May 10–15

Rapporteurs: I. Dawid, J.C. Gerhart, H.S. Horn, P.F.A. Maderson
Program Advisory Committee: J.T. Bonner (Chairman), E.H. Davidson, G.L. Freeman, S.J. Gould, H.S. Horn, G.F. Oster, H.W. Sauer, D.B. Wake, L. Wolpert
Editor: J.T. Bonner

1982. 4 photographs, 14 figures, 6 tables. X, 357 pages. ISBN 3-540-11331-2

Background papers by J.T. Bonner, R.J. Britten, E.H. Davidson, N.K. Wessells, G.L. Freeman, L. Wolpert, T.C. Kaufman, B.T. Wakimoto, M.J. Katz, S.C. Stearns, H.S. Horn, P. Alberch, S.J. Gould and group reports by numerous specialists

Physical and Chemical Sciences Research Report
Editor: S. Bernhard
Volume 3

Mineral Deposits and the Evolution of the Biosphere

Report of the Dahlem Workshop on Biospheric Evolution and Precambrian Metallogeny, Berlin 1980, September 1–5

Rapporteurs: S.M. Awramik, A. Button, J.H. Oehler, N. Williams
Program Advisory Committee: S.M. Awramik, A. Babloyantz, P. Cloud, G. Eglinton, H.L. James, C.E. Junge, I.R. Kaplan, S.L. Miller, M. Schidlowski, P.H. Trudinger
Editors: H.D. Holland, M. Schidlowski

1982. 4 photographs, 41 figures, 9 tables. X, 333 pages. ISBN 3-540-11328-2

Background papers by H.D. Holland, M. Schidlowski, H.G. Trüper, K.L.H. Edmunds, S.C. Brassell, G. Englinton, K.H. Nealson, S.M. Awramik, J. Langridge, D.M. McKirdy, J.H. Hahn, S.L. Miller, P.A. Trudinger, N. Williams, H.H.L. James, A.F. Trendall, R.E. Folinsbee, A. Lerman and group reports by numerous specialists

Springer-Verlag
Berlin
Heidelberg
New York